NATIONAL ACADEMIES *Sciences Engineering Medicine*

NATIONAL ACADEMIES PRESS
Washington, DC

Off-Lake Sources of Airborne Dust in Owens Valley, California

Owens Lake Scientific Advisory Panel

Board on Earth Sciences and Resources

Board on Environmental Studies and Toxicology

Division on Earth and Life Studies

Consensus Study Report

NATIONAL ACADEMIES PRESS 500 Fifth Street, NW Washington, DC 20001

This activity was supported by a contract between the National Academy of Sciences and Great Basin Unified Air Pollution Control District with funds from the Los Angeles Department of Water and Power. Any opinions, findings, conclusions, or recommendations expressed in this publication do not necessarily reflect the views of any organization or agency that provided support for the project.

International Standard Book Number-13: 978-0-309-72637-5
Digital Object Identifier: https://doi.org/10.17226/27958

This publication is available from the National Academies Press, 500 Fifth Street, NW, Keck 360, Washington, DC 20001; (800) 624-6242; https://nap.nationalacademies.org.

The manufacturer's authorized representative in the European Union for product safety is Authorised Rep Compliance Ltd., Ground Floor, 71 Lower Baggot Street, Dublin D02 P593 Ireland; www.arccompliance.com.

Copyright 2025 by the National Academy of Sciences. National Academies of Sciences, Engineering, and Medicine and National Academies Press and the graphical logos for each are all trademarks of the National Academy of Sciences. All rights reserved.

Printed in the United States of America.

Suggested citation: National Academies of Sciences, Engineering, and Medicine. 2025. *Off-Lake Sources of Airborne Dust in Owens Valley, California*. Washington, DC: National Academies Press. https://doi.org/10.17226/27958.

The **National Academy of Sciences** was established in 1863 by an Act of Congress, signed by President Lincoln, as a private, nongovernmental institution to advise the nation on issues related to science and technology. Members are elected by their peers for outstanding contributions to research. Dr. Marcia McNutt is president.

The **National Academy of Engineering** was established in 1964 under the charter of the National Academy of Sciences to bring the practices of engineering to advising the nation. Members are elected by their peers for extraordinary contributions to engineering. Dr. Tsu-Jae Liu is president.

The **National Academy of Medicine** (formerly the Institute of Medicine) was established in 1970 under the charter of the National Academy of Sciences to advise the nation on medical and health issues. Members are elected by their peers for distinguished contributions to medicine and health. Dr. Victor J. Dzau is president.

The three Academies work together as the **National Academies of Sciences, Engineering, and Medicine** to provide independent, objective analysis and advice to the nation and conduct other activities to solve complex problems and inform public policy decisions. The National Academies also encourage education and research, recognize outstanding contributions to knowledge, and increase public understanding in matters of science, engineering, and medicine.

Learn more about the National Academies of Sciences, Engineering, and Medicine at **www.nationalacademies.org**.

Consensus Study Reports published by the National Academies of Sciences, Engineering, and Medicine document the evidence-based consensus on the study's statement of task by an authoring committee of experts. Reports typically include findings, conclusions, and recommendations based on information gathered by the committee and the committee's deliberations. Each report has been subjected to a rigorous and independent peer-review process and it represents the position of the National Academies on the statement of task.

Proceedings published by the National Academies of Sciences, Engineering, and Medicine chronicle the presentations and discussions at a workshop, symposium, or other event convened by the National Academies. The statements and opinions contained in proceedings are those of the participants and are not endorsed by other participants, the planning committee, or the National Academies.

Rapid Expert Consultations published by the National Academies of Sciences, Engineering, and Medicine are authored by subject-matter experts on narrowly focused topics that can be supported by a body of evidence. The discussions contained in rapid expert consultations are considered those of the authors and do not contain policy recommendations. Rapid expert consultations are reviewed by the institution before release.

For information about other products and activities of the National Academies, please visit www.nationalacademies.org/about/whatwedo.

OWENS LAKE SCIENTIFIC ADVISORY PANEL

ARMISTEAD (TED) RUSSELL (*Chair*), Georgia Institute of Technology
SARAH AARONS, University of California, San Diego
ROYA BAHREINI, University of California, Riverside
DAVID DUBOIS, New Mexico State University
VALERIE EVINER, University of California, Davis
SHANNON MAHAN, U.S. Geological Survey
TOM MOORE, Regional Air Quality Council (*resigned from committee December 2024*)
GREGORY OKIN, University of California, Los Angeles
DANI OR, University of Nevada, Reno
ROBERT SCOTT VAN PELT, United States Department of Agriculture
AKULA VENKATRAM, University of California, Riverside
IAN WALKER, University of California, Santa Barbara

National Academies of Sciences, Engineering, and Medicine Staff

MARGO REGIER, Program Officer (*until May 2025*)
STEPHANIE JOHNSON, Senior Program Officer
DOMINIQUE JENKINS, Program Assistant
THOMASINA LYLES, Senior Program Assistant
MILES LANSING, Program Coordinator

BOARD ON EARTH SCIENCES AND RESOURCES

MICHAEL MANGA (*Chair*), University of California, Berkeley
MICHELE L. COOKE, University of Massachusetts, Amherst
BRADLEY CRAMER, University of Iowa
MARY FEELEY, ExxonMobil Exploration Company
YOUSSEF M. HASHASH, University of Illinois Urbana-Champaign
DOUG HOLLETT, MH Technology Partners
KATHERINE W. HUNTINGTON, University of Washington
KRISTEN KURLAND, Carnegie Mellon University
JESSICA MOORE, West Virginia Geological and Economic Survey
ANN J. OJEDA, Auburn University
DAVID SPEARS, Virginia Department of Energy
DAVE SZYMANSKI, Bentley University
JOLANTE VAN WIJK, Los Alamos National Laboratory
JESSICA M. WARREN, University of Delaware

National Academies of Sciences, Engineering, and Medicine Staff

DEBORAH GLICKSON, Board Director
LAURA EHLERS, Senior Program Officer
STEPHANIE JOHNSON, Senior Program Officer
SAMMANTHA MAGSINO, Senior Program Officer
CHARLES BURGIS, Program Officer
MARGO REGIER, Program Officer (*until May 2025*)
JONATHAN TUCKER, Program Officer
NOEL WALTERS, Associate Program Officer
MILES LANSING, Program Coordinator
MAYA FREY, Senior Program Assistant
DOMINIQUE JENKINS, Program Assistant
SAMUEL KRAFT, Program Assistant
BRYAN RUFF, Program Assistant

BOARD ON ENVIRONMENTAL STUDIES AND TOXICOLOGY

FRANK W. DAVIS (*Chair*), University of California, Santa Barbara
DANA BOYD BARR, Emory University
WEIHSUEH A. CHIU, Texas A&M University
FRANCESCA DOMINICI, Harvard University
BENIS N. EGOH, University of California, Irvine
CORIE A. ELLISON, The State University of New York at Buffalo
MAHMUD FAROOQUE, Arizona State University
MARIE C. FORTIN, Jazz Pharmaceuticals
SUSAN P. HARRISON, University of California, Davis
MARIE L. MIRANDA, University of Illinois, Chicago
MELISSA J. PERRY, George Mason University
JOSHUA TEWKSBURY, Smithsonian Tropical Research Institute
SACOBY M. WILSON, University of Maryland
TRACEY J. WOODRUFF, University of California, San Francisco

National Academies of Sciences, Engineering, and Medicine Staff

CLIFFORD S. DUKE, Board Director (*until May 2025*)
NATALIE ARMSTRONG, Program Officer
ANTHONY DEPINTO, Program Officer
KATHRYN GUYTON, Senior Program Officer
LAURA LLANOS, Finance Business Partner
THOMASINA LYLES, Senior Program Assistant

Reviewers

This Consensus Study Report was reviewed in draft form by individuals chosen for their diverse perspectives and technical expertise. The purpose of this independent review is to provide candid and critical comments that will assist the National Academies of Sciences, Engineering, and Medicine in making each published report as sound as possible and to ensure that it meets the institutional standards for quality, objectivity, evidence, and responsiveness to the study charge. The review comments and draft manuscript remain confidential to protect the integrity of the deliberative process.

We thank the following individuals for their review of this report:

SARAV ARUNACHALAM, University of North Carolina, Chapel Hill
THURE CERLING, University of Utah
JOHN DICKEY, PlanTierra LLC
AMY EAST, U.S. Geological Survey
MICHAEL FLAGG, Bay Area Air Quality Management District
STEVE HANNA, Hanna Consultants
NICK LANCASTER, Desert Research Institute
DANIEL McEVOY, Desert Research Institute
THERESA NAJITA, California Air Resources Board
SUJITH RAVI, Temple University
SCOTT TYLER, University of Nevada, Reno
AMY ZIMPFER, U.S. Environmental Protection Agency (*retired*)

Although the reviewers listed above provided many constructive comments and suggestions, they were not asked to endorse the conclusions or recommendations of this report nor did they see the final draft before its release. The review of this report was overseen by **DAVID ALLEN (NAE),** University of Texas at Austin, and **MARK SWEENEY,** University of South Dakota. They were responsible for making certain that an independent examination of this report was carried out in accordance with the standards of the National Academies and that all review comments were carefully considered. Responsibility for the final content rests entirely with the authoring committee and the National Academies.

Acknowledgments

The Owens Lake Scientific Advisory Panel (OLSAP) would like to thank the following people who gave presentations, participated in panel discussions, and provided public comments.

Arrash Agahi, Los Angeles Department of Water and Power
Steve Bacon, Desert Research Institute
Kathy Bancroft, Lone Pine Paiute Shoshone Reservation
John Bannister, Air Sciences
Nik Barbieri, Great Basin Unified Air Pollution Control District
Alex Clayton, Great Basin Unified Air Pollution Control District
Stella Cook, U.S. Environmental Protection Agency
Wes Danskin, U.S. Geological Survey
Hank Dickey, Consultant
Scott Epstein, South Coast Air Quality Monitoring District
Sondra Grimm, Great Basin Unified Air Pollution Control District
Alex Hall, University of California, Los Angeles
Yousaf Hammed, Clark County Department of Environment and Sustainability
Victor Harris, Stantec Consulting
Grace Holder, Great Basin Unified Air Pollution Control District
Shaye Hong, U.S. Environmental Protection Agency
Mel Joseph, Lone Pine Paiute Shoshone Tribe
Phillip Kiddoo, Great Basin Unified Air Pollution Control District
Alicia Kindred, California Air Resources Board
Katie Kolesar, Air Sciences
Jevon Lam, Los Angeles Department of Water and Power
Ann Logan, Great Basin Unified Air Pollution Control District
Carrie MacDougall, Formation Environmental
Sally Manning, Big Pine Paiute Tribe
Leah Matthews, California Air Resources Board
Gobeail McKinley, U.S. Environmental Protection Agency

Kim Mitchell, Great Basin Unified Air Pollution Control District
Theresa Najita, California Air Resources Board
Jeff Nordin, Los Angeles Department of Water and Power
Beth Palma, U.S. Environmental Protection Agency
Michael Prather, Eastern Sierra Audubon
James Richards, University of California, Davis
Mark Schaaf, Air Sciences
Brian Schmid, Formation Environmental
Sean Scruggs, Fort Independence Indian Reservation
Jason Smesrud, Jacobs Engineering
Monica Soucier, Imperial County Air Pollution Control District
Sheila Tsai, U.S. Environmental Protection Agency
Ginger Vagenas, U.S. Environmental Protection Agency
Jaime Valenzuela, Los Angeles Department of Water and Power
Dena Vallano, U.S. Environmental Protection Agency
Jaime Lopez Wolters, Friends of the Inyo
Yohannes Yimam, Formation Environmental

Contents

SUMMARY   1

1 INTRODUCTION   9
Statement of Task and Approach, 11
Outline of Report, 13

2 A SYSTEMS-LEVEL UNDERSTANDING FOR OFF-LAKE PM_{10} SOURCES   16
Overview of Controls Over Owens Valley PM_{10} Emissions, 16
Geological Context, 16
Climatology and Hydrology, 21
Vegetation Dynamics, 27
Anthropogenic Disturbance of Vegetation and the Soil Surface, 29
Climate Change, 29
Tribal Knowledge and Cultural Priorities, 30
Summary, 31

3 SOURCES OF PM_{10} EMISSIONS IN THE OWENS VALLEY PLANNING AREA   32
Trends in Owens Lake Air Quality Over Time, 33
Current Tools and Methods for PM_{10} Source Attribution, 35
Impacts of Local Off-Lake Sources to PM_{10} Exceedances, 38
Temporal Emission Trends, 54
Additional Monitoring and Modeling Approaches, 56
Conclusions and Recommendation, 64

4 ORIGIN AND EVOLUTION OF LOCAL OFF-LAKE SOURCES IN OWENS VALLEY   66
Winnowing Hypothesis for Off-Lake Dust, 66
Northeast Side of the Lake, 67
Southern Side of the Lake, 87
North of the Lake, 92

Conclusions and Recommendations, 97
Potential Sources North of the Lake, 98
Further Research to Establish the Origin and Evolution of OVPA Sources, 99

5 UTILIZATION OF THE U.S. EPA EXCEPTIONAL EVENTS RULE IN THE OVPA 100
The Exceptional Events Rule, 100
Applicability of Exceptional Events Rule for Off-Lake Sources in the OVPA, 102
Public Health Impacts and a Reasoned Use of the Exceptional Events Rule, 105
Conclusions and Recommendation, 106

6 INFORMING DUST CONTROL DECISIONS FOR OFF-LAKE SOURCES 107
Potential Dust Control Measures on Sand Sheets and Dunes, 108
Potential Dust Control on Flood Deposits, 117
Other Potential Dust Control Measures, 120
Application of Tribal Knowledge for Dust Control, 121
Conclusions, 121

REFERENCES 126

APPENDICES

A EMISSION FLUXES OF ON- AND OFF-LAKE SOURCES 139
B ADVISORY PANEL BIOGRAPHICAL SKETCHES 143
C GLOSSARY 147

Acronyms and Abbreviations

AERONET	Aerosol Robotic Network
AGL	Above Ground Level
AOD	Aerosol Optical Depth
AQS	Air Quality System
ARS	Agricultural Research Service
BACM	best available control measures
BLM	Bureau of Land Management
CAA	Clean Air Act
Caltrans	California Department of Transportation
CARB	California Air Resources Board
CASAC	Clean Air Science Advisory Committee
CMAQ	Community Multiscale Air Quality
DEM	digital elevation model
DCM	dust control measure
DRI	Desert Research Institute
EER	Exceptional Events Rule
EMIT	Earth Surface Mineral Dust Source Investigation
EPA	Environmental Protection Agency
FEM	Federal Equivalent Method
FRM	Federal Reference Method
GBUAPCD	Great Basin Unified Air Pollution Control District
GDV	groundwater-dependent vegetation
GOES	Geostationary Operational Environmental Satellites

ICPMS	inductively coupled plasma mass spectrometry
LA	Los Angeles
LADWP	Los Angeles Department of Water and Power
MAIA	Multi-Angle Imager for Aerosols
MODIS	Moderate Resolution Imaging Spectroradiometer
mph	miles per hour
NAAQS	National Ambient Air Quality Standards
NASA	National Aeronautics and Space Administration
NASEM	National Academies of Science, Engineering, and Medicine
NOAA	National Oceanic and Atmospheric Administration
OHV	off-highway vehicle
OLSAP	Owens Lake Scientific Advisory Panel
OSL	optically stimulated luminescence
OVPA	Owens Valley Planning Area
PI-SWERL	Portable In Situ Wind Erosion Laboratory
PM	Particulate Matter
POC	Parameter Occurrence Code
RACM	reasonably available control measures
RMSE	root mean squared error
SIP	State Implementation Plan
SSA	singular spectrum analysis
TEOM	tapered element oscillating microbalance
UV	ultraviolet
VIIRS	Visual Infrared Imaging Radiometer Suite
WRF	Weather Research and Forecasting
XANES	X-ray absorption near edge structure

Summary

Owens Lake (*Patsiata*) is located at the southern end of the Owens Valley (*Payahuunadü*) in California. After water was diverted into the Los Angeles Aqueduct starting in 1913, the dry lakebed became one of the largest sources of airborne particulate matter in the United States, including particulate matter with an aerodynamic diameter of 10 micrometers or less (PM_{10}). These small particles can penetrate into the lungs and cause or worsen a variety of health problems like asthma, bronchitis, chronic obstructive pulmonary disease, and respiratory infections. Since 2001, the Los Angeles Department of Water and Power (LADWP), at the direction of the Great Basin Unified Air Pollution Control District (which this report will refer to as "the District"), has implemented dust control measures on the lakebed. These on-lake sources are defined as those that lie below the 3,600-ft elevation regulatory shoreline that defines the Owens Lake bed (Figure S-1). The District has also implemented approximately 140 acres of dust control measures on off-lake areas (above the 3,600-ft regulatory shoreline) at the Keeler Dunes. Together, these controls have made substantial progress toward reducing the frequency and intensity of exceedances of the PM_{10} standard. However, both on-lake and off-lake sources continue to cause PM_{10} exceedances that prevent the Owens Valley Planning Area (OVPA) from attaining the PM_{10} National Ambient Air Quality Standard (NAAQS).

The National Academies of Sciences, Engineering, and Medicine's Owens Lake Scientific Advisory Panel (OLSAP) was established following a 2014 stipulated judgment between the District and LADWP to foster understanding and collaboration on the scientific and technical approaches to dust control in the OVPA. OLSAP's first task focused on the effectiveness of on-lake dust control measures, and they released a report on the topic in 2020 titled *Effectiveness and Impacts of Dust Control Measures for Owens Lake*. For its second task, the panel was asked to summarize the impact of off-lake sources on PM_{10} exceedances, as well as their distribution, origin, and how they might change over time. The panel was asked to recommend better methods for characterizing and monitoring PM_{10} contributions by off-lake sources. Additionally, the panel was asked to discuss possible dust control measures that could be applied to off-lake sources, as well as the applicability of the U.S. Environmental Protection Agency's (EPA's) Exceptional Events Rule for excluding air quality monitoring data from unusual or naturally occurring events that are not reasonably controllable. To gather the necessary information to address the statement of task, the panel held several information-gathering sessions, including a 2-day meeting at Owens Lake.

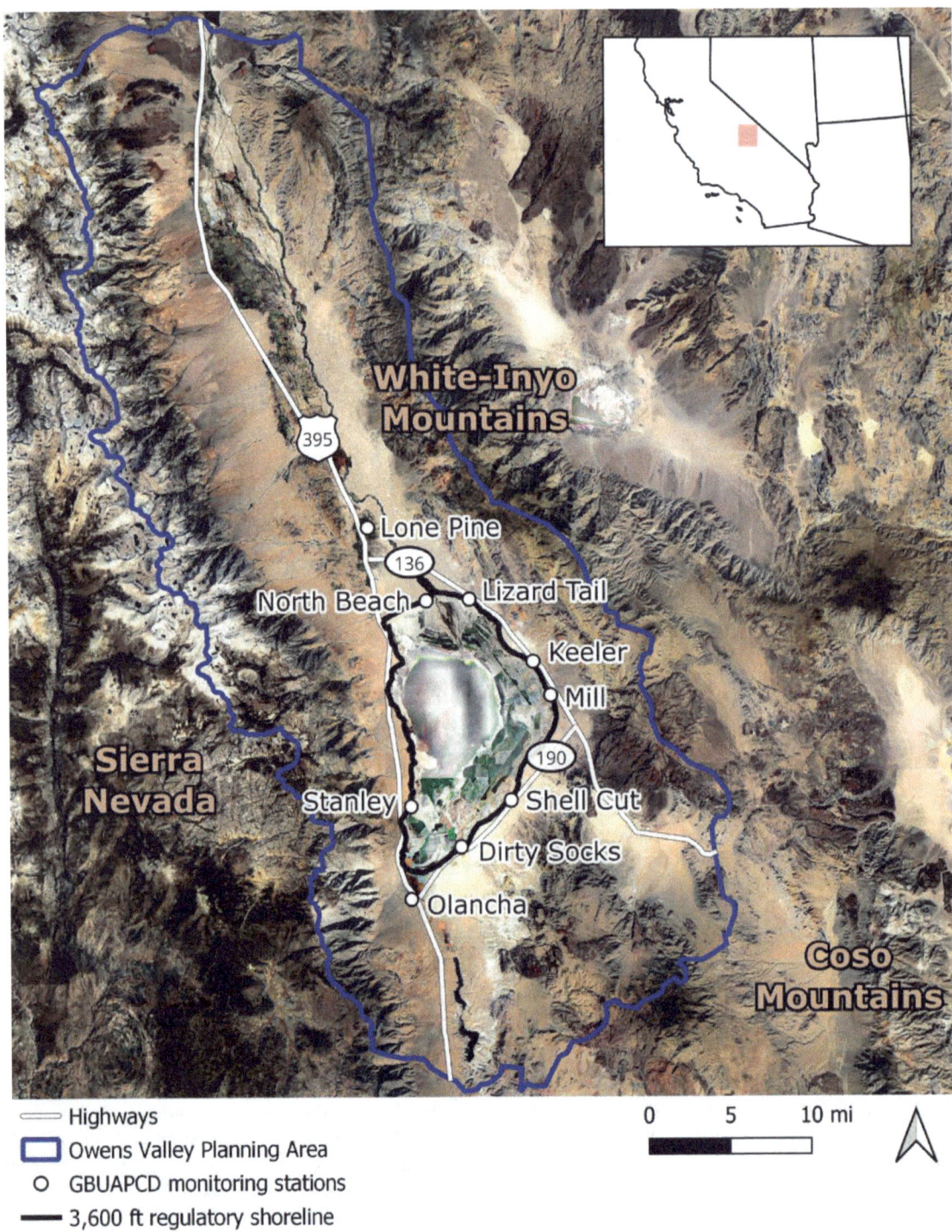

FIGURE S-1 The Owens Valley Planning Area (OVPA) and the 3,600-ft regulatory shoreline, which defines the Owens Lake bed.
NOTES: PM_{10} monitoring stations in the OVPA that are operated by Great Basin Unified Air Pollution Control District (GBUAPCD) are indicated, as well as surrounding mountain ranges.

The panel found that off-lake sources are, and will remain, a major source of PM_{10} exceedances in the OVPA. Local off-lake sources in the OVPA include flood deposits,[1] Keeler Dunes, Olancha Dunes, alluvial fans, up-valley sources, and anthropogenic disturbances, among others. PM_{10} emissions from some of these sources, like certain alluvial fans and flood deposits not impacted by highway infrastructure, are judged with high certainty to be natural in origin. It would be reasonable to apply the Exceptional Events Rule to applicable high-wind events from these naturally occurring and not-reasonably controllable sources so that these data would be excluded from regulatory decisions. Exceedances from other sources—Keeler Dunes, Olancha Dunes, and flood deposits influenced by highway infrastructure—are at least partially anthropogenic in origin and thus dust control measures may be required. Further monitoring and modeling would define more clearly the sources of PM_{10} emissions and their natural or anthropogenic origin. If dust control measures are required, the panel found that the establishment of native vegetation would be the most stable and sustainable dust control measure across all off-lake surfaces emitting particulate matter. On sand sheets and dunes, natural roughness features such as straw bales or straw checkerboards would be necessary components of revegetation efforts. On small, impounded flood deposits, gravel cover could provide near-term dust control, but longer-term control would require improved drainage infrastructure to reduce accumulation of fine-grained sediments. In all cases, early engagement of local Tribal Nations in the planning and design of dust control methods and consideration of the resilience of these dust control methods under a changing climate are essential for reducing PM_{10} emissions in the OVPA.

SOURCES OF PM_{10} EMISSIONS IN THE OWENS VALLEY PLANNING AREA

Dust control measures have made substantial progress toward reducing the frequency and intensity of on-lake exceedances, but both on-lake and off-lake sources continue to cause PM_{10} exceedances in the OVPA. The relatively consistent number of PM_{10} exceedances that the District has attributed to off-lake sources over the last 25 years, despite trends indicating a declining number of exceedances from on-lake sources, demonstrates the importance of these off-lake sources and suggests that these sources could hinder attainment with the PM_{10} NAAQS in the region.

Conclusion 3-1: Off-lake sources currently contribute the majority of exceedances of the PM_{10} NAAQS at most monitoring sites in the OVPA and are likely to remain important contributors in the future.

Since 2017, the District has used additional information to attribute PM_{10} exceedances to specific sources within the OVPA. This information includes particulate and meteorological data, modeling, cameras, field observations, and media reports of dust storms. These data are compiled into the District's exceedance database. Based on these data, the District classifies each exceedance as one of the following: 1) dust—primarily on-lake sources, 2) dust—primarily local off-lake sources, 3) dust—primarily regional event, 4) wildfire smoke, or 5) mixed—dust and wildfire sources. Each documented exceedance includes detailed comments on likely source areas. The panel supports the District's general approach to source apportionment and used this information to identify a few specific local off-lake sources that cause a disproportionate impact on the PM_{10} exceedances in the OVPA. These include flood deposits (including channelized, sheet/overland flow and impounded flood deposits), Keeler Dunes, Olancha Dunes, alluvial fans, up-valley sources, and anthropogenic disturbances.

Conclusion 3-2: The most frequent local off-lake source of exceedances from 2017 to 2024 is flood deposits, followed by Olancha Dunes and Keeler Dunes.

This District's method for source attribution is useful for assessing broad trends, but the classifications are nonquantitative, and some uncertainties remain in the identification of specific off-lake source areas. For example, the current methodology does not allow for the quantification of PM_{10} contributions from different sources for a

---

[1] Following the District's methods for source attribution, "flood deposits" in this report is defined as including 1) channelized flood deposits; 2) sheet/overland flow deposits; and 3) deposits from impounded floodwaters. The term "alluvial fans," in contrast, is used to refer to a general landscape feature and not a specific flood event.

single exceedance, and data are often not collected in locations that are ideal to capture detailed information about off-lake sources. Additional measurements and modeling would enable the District to more definitively identify how specific sources have contributed to exceedances and support future air quality management decisions.

Recommendation 3-1: Given the importance of better characterizing contributing sources to individual exceedances from off-lake sources, the District, the California Air Resource Board (CARB), the U.S. Environmental Protection Agency (EPA) and land owners/managers should consider supporting the following measurements and modeling:

- **Compositional analysis (species, elements, isotopes) of PM_{10} material to better identify source areas leading or significantly contributing to exceedances;**
- **Additional cameras to better attribute specific off-lake sources;**
- **Portable In-Situ Wind Erosion Lab (PI-SWERL) transects to identify changes to dust emission potential with increased human influence, such as at the Olancha Dunes Off-Highway Vehicle Area;**
- **Temporary monitoring of PM_{10} to better characterize source areas of concern, potentially including the Olancha Dunes Off-Highway Vehicle Recreational Area and fallow agricultural fields and former groundwater-dependent vegetation areas to the north of the lake;**
- **PM_{10} modeling of off-lake sources, potentially using AERMOD, an EPA-recommended dispersion model for local-scale modeling instead of, or alongside, CALPUFF.**

ORIGIN AND EVOLUTION OF OFF-LAKE SOURCES IN OWENS VALLEY

The panel considered multiple lines of evidence to infer the origins and evolution of major PM_{10} sources in the OVPA, or those that might become important sources in the future. One important process the panel considered was winnowing. This process suggested that PM_{10} material from the dry lakebed deposited onto off-lake landforms would be expected to decrease over time due to its resuspension and removal by aeolian processes. This hypothesis was based on a correlation between on- and off-lake exceedances at the Dirty Socks monitor between 1999 and 2012, which has not continued in more recent estimated emissions trends and exceedance data. Most current PM_{10} emissions from off-lake areas are likely not a result of resuspension of PM_{10} material originating from the lakebed that was deposited on off-lake landforms. Instead, the presence and common replenishment of highly emissive flood deposits provide ample fine particulates that can be emitted as PM_{10} as long as the horizontal flux of saltation-sized particles is sufficient to emit dust from the surface.

Conclusion 4-1: Winnowing is expected to play a minimal role in reduction in future off-lake PM_{10} exceedances.

Northeast Side of the Lake

The northeastern side of Owens Lake is host to several landforms including the Keeler Dunes and the Slate Canyon/Keeler Alluvial Fan Complex that have undergone major changes over the last century and are substantial contributors to exceedances. During the 20th century, the Keeler Dunes transitioned from a largely vegetated dune system that was stabilized by greasewood (*Sarcobatus vermiculatus*) to an active dune field. The emergence of Keeler Dunes as an active dune field resulted in abundant saltation that can drive PM_{10} emissions from flood deposits that are continually replenished from the alluvial fan. Therefore, ongoing PM_{10} exceedances from this area are a direct result of the destabilization of the Keeler Dunes. The panel finds that the net transport direction, available imagery, and the evidence for groundwater-dependent vegetation currently present in the dunes support the conclusion that increased sand transport following the diversion of water from Owens Lake destabilized the Keeler Dunes. Changes to surface hydrology resulting from construction of berms designed to protect Highway 136 appear to have had an impact on upland (non-groundwater-dependent) vegetation but are unlikely to have led to the destabilization of groundwater-dependent vegetation in the Keeler Dunes.

Conclusion 4-2: The reactivation of the Keeler Dunes was related to the additional upwind sand supply available from the Owens River delta following drainage of Owens Lake.

Conclusion 4-3: Due to continuing aeolian activity of the Keeler Dunes and replenishment of flood deposits within the dunes, the system will continue to contribute material to PM_{10} emissions. Stabilization of the dunes would likely reduce PM_{10} emissions.

Several constructed berms northeast of Highway 136 were intentionally designed to alter surface hydrology, directing overland flow to specific discharge points along the highways. The panel did not analyze each berm-related flood deposit but instead considered the berms on the Slate Canyon/Keeler Alluvial Fan Complex as potentially representative of similar features around Owens Lake. These berms have had appreciable, localized impacts on the distribution of flood deposits in the Keeler Dunes region, especially following impacts from the remnants of Hurricane Kay in September of 2022. Further investigation would be needed to determine the impacts that these berms have, if any, on potential off-lake PM_{10} emissions from flood deposits.

Conclusion 4-4: The construction of berms northeast of Keeler Dunes and elsewhere modified sediment transport, but it is uncertain if this modification of sediment transport increased PM_{10} emissions from flood deposits relative to that which would have occurred without berms.

Recommendation 4-1: The District should work with the California Department of Transportation and other Owens Valley Planning Area landowners to determine the impact of berms on flood deposits and associated PM_{10} emissions.

Southern Side of the Lake

The southern side of Owens Lake is host to the Olancha Dunes and multiple alluvial systems that are important sources of PM_{10} exceedances. The scientific literature on the origin and evolution of the Olancha Dunes is quite sparse, but available evidence indicates that the dunes formed prior to the diversion of water from Owens Lake. The panel's analysis shows that the dunes experienced a slight southward extension (approximately 0.3 miles) from 1944 to the current day. This southward extension could be the result of increased sediment supply following the diversion of water from Owens Lake or from other natural or anthropogenic activities.

Olancha Dunes is also the location of an off-highway vehicle (OHV) and dispersed camping recreational area that makes up approximately 36 percent of the total dune area. There is little to no research on the impacts of decades of OHV activity on PM_{10} emissions at Olancha Dunes, but research from other sites like the Oceano Dunes State Vehicular Recreation Area and the Imperial Sand Dunes Recreation Area shows clear associations between OHV activity, decreasing vegetation cover, and increased dust emissions. Additional study using aerial photography, PI-SWERL dust emission potential measurements, and the Bureau of Land Management's records of impacts from recreational activity could provide information on the contribution of recreational activity to PM_{10} emissions.

Conclusion 4-5: The Olancha Dunes has extended southward slightly since the 1940s, but there is not sufficient evidence to indicate that this southward extension was influenced by drainage of the lake or other anthropogenic activities, such as OHV recreation and dispersed camping.

Recommendation 4-2: The District should work with the Bureau of Land Management to determine the impacts of recreational activity on plant communities and PM_{10} emissions within the Olancha Dunes and remediate as needed.

The southern side of Owens Lake also hosts multiple alluvial channel/wash systems that deliver and rework sediments from the neighboring Coso and Sierra Nevada ranges. These alluvial channel/wash systems supply sand

and PM_{10} material, and they can only support low-density vegetation cover, which creates conditions ripe for high PM_{10} emission. While the replenishment of these alluvial systems is a natural process that has been occurring for millennia, anthropogenic alteration of the flowpaths through the constructed infrastructure may change the amount and distribution of impounded water and sediment and thereby change its potential to contribute to PM_{10} emissions. Climate change is projected to make extreme precipitation events more frequent and intense, which would more frequently replenish fine sediments in flood deposits that contribute to PM_{10} emissions.

Conclusion 4-6: Aerial and satellite images suggest that the impounded flood deposits south of the lake near the Dirty Socks PM_{10} monitor may have been affected by the rerouting of Highway 190. Highway 190 infrastructure clearly impacts flood flows in other areas along the south of the lake, although it is unclear to what extent, if any, this infrastructure impacts overall PM_{10} emissions and measured exceedances.

Recommendation 4-3: The District should work with the California Department of Transportation to determine the impact of Highway 190 and related berms on flood deposits and associated PM_{10} emissions, with initial emphasis on the impounded flood deposits near the Dirty Socks PM_{10} monitor.

Potential Sources North of the Lake

Current data indicate a stable shallow groundwater table in the area around Owens Lake. However, there is substantial evidence that areas north of the lake have seen decreases in vegetation cover, which may have contributed to historical dust emission in the OVPA. A groundwater management plan and a number of revegetation projects were implemented in the 1990s to reduce blowing dust in affected areas. If this land is not managed carefully, dust emission from the area north of the lake could increase, especially under changing climate conditions.

Conclusion 4-7: Drought coupled with constant or increasing water extraction in the Owens Valley could result in prolonged lowering of the groundwater table. If groundwater drops to levels that severely impact the health of existing groundwater-dependent vegetation, the potential for PM_{10} emissions north of the lake would increase.

Conclusion 4-8: Continued monitoring and regular updates on the advancement of revegetation projects on former groundwater-dependent meadows and abandoned agricultural fields will inform potential measures that may be necessary to reduce PM_{10} emissions in the face of future climate pressures.

Further Research to Establish the Origin and Evolution of OVPA Sources

More chronological research may reduce uncertainties surrounding the origin and evolution of dune fields and flood deposits. Collecting sediment samples across dunes and flood deposits by coring or auguring may be the best way to collect a comprehensive set of data. These methods may illuminate relatively recent processes that occurred after the diversion of water from Owens Lake, construction of berms, and the rerouting of Highway 190.

Conclusion 4-9: A coring and optically stimulated luminescence campaign targeting recent mobilization events (including those younger than 100 years) across Olancha Dunes, Keeler Dunes, and the flood deposits near the Dirty Socks monitor will reduce uncertainty on the origin and evolution of these deposits.

UTILIZATION OF THE EXCEPTIONAL EVENTS RULE IN THE OVPA

The EPA's Exceptional Events Rule for high wind dust events can be used to exclude unusual or naturally occurring exceedance event data from consideration in regulatorily significant decisions, as long as other public health protections are met. Nevertheless, the way the Exceptional Events Rule is interpreted and implemented has consequences for a region's air quality and associated health impacts. Based on current EPA guidance and the panel's evaluation of the origin and evolution of off-lake sources, some exceedances from off-lake sources in the OVPA may be considered natural events, as human activity played little or no direct causal role in their occurrences.

Conclusion 5-1: Most exceedances from channelized flow deposits, sheet flow deposits, and deposits impounded behind natural features appear to fit the Exceptional Events Rule criterion of natural events.

Some emissions from off-lake sources in the OVPA may be considered anthropogenically influenced. Since the OVPA has a State Implementation Plan that is over 5 years old, the Exceptional Events Rule requires an assessment and implementation of reasonable controls that could be applied to these anthropogenically influenced sources. It is beyond the charge to the panel to determine what controls are "reasonable" because that includes policy judgments rather than purely scientific assessments.

Conclusion 5-2: Emissions from the Keeler Dunes, the Olancha Dunes, and the highway-impounded flood deposit south of the Dirty Socks monitor have been affected by human activities resulting from draining of the lake, OHV recreation, and highway construction, respectively. Therefore, an assessment and potential implementation of reasonable controls would be required before an Exceptional Event demonstration is considered at these locations.

A thorough consideration of reasonable controls would include attention to resilience of the dust control measure under climate change. These considerations may include accounting for more intense flooding and prolonged drought.

Recommendation 5-1: The District, California Air Resources Board, and the Environmental Protection Agency should consider resilience under climate change as part of its assessment of reasonable controls.

INFORMING DUST CONTROL DECISIONS FOR OFF-LAKE SOURCES

If dust control measures are determined to be necessary and feasible for off-lake sources, implementation will require a systems-level landscape approach that considers cultural resources. Collaboration with local Tribal Nations will improve community investment in restoration efforts and outcomes.

Conclusion 6-1: Tribal input into the evaluation of potential dust control measures, starting at the very initial stages of project conceptualization and design, will support collaborative planning, community engagement, and successful implementation.

Many areas around the OVPA are extremely dynamic settings, requiring different approaches over space and possible re-treatment over time (e.g., in flood deposits). Nevertheless, the panel found that the establishment and maintenance of vegetation offers the best chance for a natural, self-sustaining protection of the soil surface from wind. Additional methods that will support the eventual establishment of vegetation are expanded upon below.

Conclusion 6-2: Establishing and maintaining native vegetation is the most stable and sustainable dust control measure across all emitting off-lake surfaces.

A number of sand sheets and dune fields (e.g., Keeler and Olancha) are distributed along the eastern and southern shorelines. Efforts to partially stabilize Keeler Dunes using solid natural roughness elements of straw bales have resulted in the successful establishment of native shrub seedlings that are providing seed for additional colonization of the sand. Porous naturally sourced roughness elements such as straw checkerboards have also been used with great success in China to protect highways, rail lines, and villages from encroaching sand. These barriers are inexpensive to build and, although relatively short-lived, may be repaired or renewed as necessary until vegetation has successfully colonized the area.

Conclusion 6-3: In sand sheets and dune fields, solid naturally sourced roughness elements like straw bales and porous natural roughness elements like straw checkerboards are effective, ecologically favorable, and potentially feasible means to provide temporary surface stabilization until native shrub communities become well established.

Fine-grained flood deposits are scattered in topographic lows along the shoreline and within the sand sheets and dunes. These fine-grained deposits are extremely emissive and efforts to control these dust sources could reduce exceedances. Where current deposits of fine-grained material are small in size, they could be covered with unlined layers of gravel or cobbles. This system would allow for rapid infiltration of water into the flood deposits, which hold water very effectively. Such constructions would allow plants to find a hospitable root zone that would provide water and nutrients for growth and reproduction, further stabilizing the surface. Longevity of this dust control measure would depend on the interval between floods and would potentially need to be renewed following a major flood event barring changes to infrastructure or local topography.

Conclusion 6-4: For near-term mitigation of the highly emissive highway-impounded flood deposits, a feasible dust control measure is covering fine-grained flood deposits with gravel or cobbles in parallel with vegetation restoration.

Conclusion 6-5: The panel could not identify any long-term, cost-effective dust control measures that could stabilize the large-scale flood channel deposits deposited downgradient of the berm near Keeler Dunes by the remnants of Hurricane Kay in 2022.

Flood deposits around Owens Lake have been affected by both the construction of the highway and a number of upgradient berms. Drainage improvements along the highway with the addition of culverts or the elevation of roadways would reduce future ponding of floodwater and accumulation of material next to the highway leading to PM_{10} emissions. Highway berms concentrate flow toward a limited number of culverts under the highway, which may contribute to large deposits of fine-grained material after major precipitation events (e.g., the remnants of Hurricane Kay in 2022). Modification of the berm structures, highway culverts, and water harvesting and spreading are large, intensive dust management options, but they have the potential to reduce floodwater velocity and the concentration of fine-grained sediment while enhancing water storage for vegetation establishment. Water harvesting from surface runoff and water spreading from drainage features could be used to encourage shrub growth in the future when climate change may make rainfall less predictable. Increased infiltration along the upper positions of the slope might also augment groundwater elevations such that the capillary fringe might be contacted by established shrub roots. Such a system may also limit erosion of gullies and ravines by reducing the velocity of floodwater.

Conclusion 6-6: Improved drainage for flood deposits impounded behind Highway 190 would reduce accumulation of fine-grained sediments from future flood events.

Conclusion 6-7: Hydrologic modifications of the berm structures, potentially combined with improved highway drainage and upgradient water harvesting and spreading, could reduce the size of future dust sources around Owens Lake.

Additionally, OHV recreation is known to be associated with landscape changes that lead to PM_{10} exceedances. In the Owens Valley, the Bureau of Land Management manages an OHV recreation and primitive camping area of approximately 400 acres (1.6 km^2) or 36 percent of the total area of the Olancha Dunes. Additionally, there have been reports of vehicle recreation on dry backwater lakes causing dust emissions.

Conclusion 6-8: Limits to recreational use, including off-highway vehicles, is a feasible dust control measure for recreational areas that contribute to PM_{10} exceedances.

1

Introduction

Owens Lake, or *Patsiata* in the local Mono language, is located at the southern end of the Owens Valley and bordered by the Sierra Nevada Mountains to the west, the White-Inyo Mountains to the east, and the Coso Range to the south and east (Figure 1-1). Prior to the 20th century, the lake was a closed-basin saline lake of approximately 100 square miles (Holder 1997; NASEM 2020). The lake was and still is central to the way of life for the *Nüümü* and *Newe* (Paiute and Shoshone in English, respectively) who live in Owens Valley, or *Payahuunadü*, land of the flowing water (Owens Valley Indian Water Commission 2024). Beginning in 1913, water was diverted from the Owens River into the Los Angeles Aqueduct. This diversion eventually shrank Owens Lake to one-third of its former area, leaving a brine pool surrounded by a dry lakebed (NASEM 2020). Once exposed, the lakebed produced large amounts of airborne dust under high winds. This resulted in Owens Lake being one of the largest sources of airborne particulate matter in the United States (EPA 2017; Ono 2006), including particulate matter with an aerodynamic diameter of 10 micrometers or less (PM_{10}). These small particles can penetrate the lungs and can cause or worsen a variety of health problems such as asthma, bronchitis, chronic obstructive pulmonary disease, and respiratory infections (GBUAPCD 2016). In 1987, the U.S. Environmental Protection Agency (EPA) identified the southern Owens Valley (known as the Owens Valley Planning Area, or OVPA) as violating the health-based PM_{10} National Ambient Air Quality Standard (NAAQS) (GBUAPCD 2016). The area also has been designated by the state of California as being in nonattainment of the corresponding state standards, the California Ambient Air Quality Standard (CAAQS) (NASEM 2020).[1]

Over the last 25 years, the Los Angeles Department of Water and Power (LADWP), at the direction of the Great Basin Unified Air Pollution Control District (the District), has implemented dust control measures on the lakebed, defined as the land located below the regulatory shoreline, which was designated at an elevation of 3,600 ft (GBUAPCD 2016). These dust control methods were implemented with the objective of meeting the EPA NAAQS for PM_{10}, as well as the PM_{10} standards set by the state of California. There have been four approved final State Implementation Plans (SIPs)—1998, 2003, 2008, 2016—and an amendment in 2013, that documented how PM_{10} emissions would be reduced to comply with the NAAQS (GBUAPCD 1998, 2003, 2008, 2013, 2016). As of this publication, LADWP has spent approximately $2.5 billion on their dust mitigation program at Owens Lake and implemented controls on 49.18 square miles (including 1.21 square miles of deferred areas) with a

---

[1] Attainment of the National Ambient Air Quality Standard (NAAQS) has precedence over attainment of the California Ambient Air Quality Standard (CAAQS). Unlike NAAQS, CAAQS does not require that standards be met by specified dates. Instead, the law requires incremental progress toward attainment (CARB 2024).

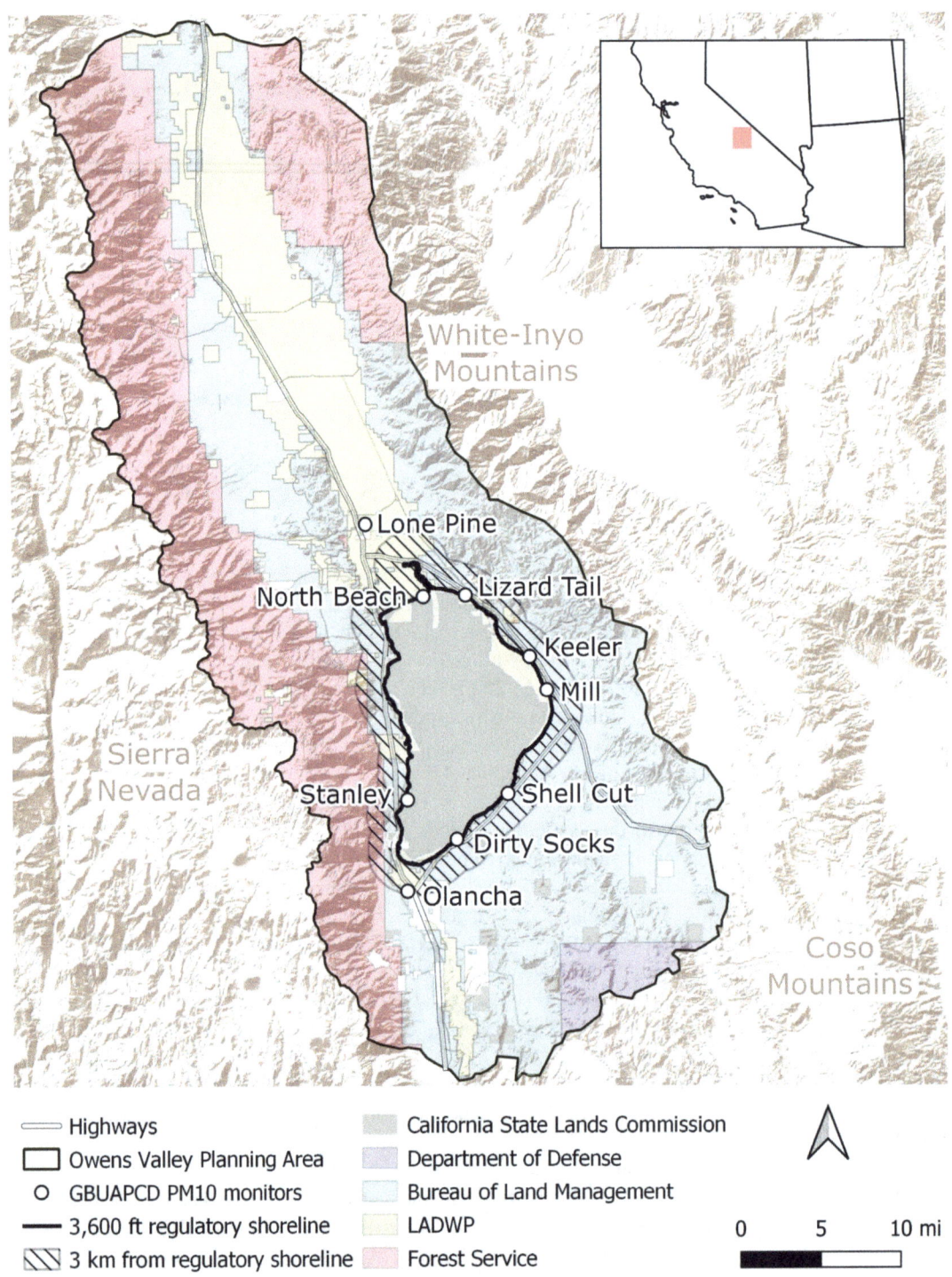

FIGURE 1-1 Map of the Owens Valley Planning Area, showing major landowners, PM_{10} monitoring stations, and mountain ranges.
NOTES: Hatch-markings delineate the area within 3 km of the 3,600-ft-elevation regulatory shoreline that defines the Owens Lake bed. Three kilometers are shown here based on the panel's statement of task, which states, "the majority of the local off-lake sources fall within 2–3 km from the shoreline...."

INTRODUCTION

potential maximum buildout extent of 53.4 square miles (Figure 1-2) (LADWP 2024b). These controls will be maintained in perpetuity. The District has also implemented approximately 140 acres of dust control measures on off-lake areas (above the 3,600-ft-elevation regulatory shoreline) at the Keeler Dunes following a $10 million public benefit contribution from LADWP and $2.6 million from District funds (GBUAPCD 2013) (Grace Holder, GBUAPCD, personal communication, May 2024). Together, these controls have made substantial progress toward reducing the frequency and intensity of exceedances of the PM_{10} standard, currently monitored at 9 locations surrounding the lakebed (Figures 1-2 and 1-3). Despite these efforts, both on-lake and off-lake sources continue to cause PM_{10} exceedances.

STATEMENT OF TASK AND APPROACH

The Owens Lake Scientific Advisory Panel (OLSAP) was established following the 2014 stipulated judgment between the District and LADWP. The goal of the panel is to foster understanding and collaboration on the scientific and technical approaches to dust control. The panel's first report (NASEM 2020) focused on the effectiveness of alternative dust control methods for on-lake sources of PM_{10}. Since that report, the OVPA has seen continued exceedances that have prevented attainment with the PM_{10} NAAQS, which requires no more than one expected exceedance per year at each monitor, averaged over 3 years (40 C.F.R. § 50.6[a] 2023) (see also Chapter 3). As shown in Figure 1-3, these continued exceedances are increasingly dominated by off-lake sources. In 2022, the District sent a letter to LADWP and the National Academies of Sciences, Engineering, and Medicine requesting that OLSAP examine and evaluate PM_{10} emissions from off-lake sources. The full statement of task is found in Box 1-1.

Given the focus of the statement of task on the PM_{10} NAAQS and not the CAAQS, the panel focused much of its analysis on attainment of the NAAQS. However, the panel recognizes that attainment of the CAAQS is an eventual goal. Additionally, while both on-lake and off-lake sources can and do contribute to varying degrees to PM_{10} exceedances on individual monitored exceedance days, the statement of task solely focuses on off-lake sources (i.e., those that are above the 3,600-ft-elevation regulatory shoreline). Although the District emphasized that the primary focus of the statement of task is local off-lake sources that are primarily within 2–3 km of Owens Lake (Table 1-1),[2] the panel also considered other sources in the Owens Valley Planning Area "for which available data indicate [as] potentially significant contributors to high PM_{10} concentrations in the Owens Valley Planning Area (OVPA)." Additionally, the panel considered potential future sources as the task mandates that the panel consider "changes in off-lake dust sources over time" (Box 1-1). The District also noted that windblown sources of dust are the focus of the task, not local permitted sources of air pollution (e.g., landfills, mines) nor other local sources of dust from mechanical and combustion processes (e.g., woodburning devices, transportation, unpaved roads, paved road dust, tailpipe emissions, prescribed burning) (Table 1-1). Similarly, regional dust events originating from outside of the OVPA and wildfire smoke events are not the main focus of the task, although these ancillary sources of PM_{10} may be relevant to the discussion, as they can contribute to exceedances in the OVPA (Table 1-1). Finally, as suggested by the statement of task, the panel only considered the potential application of the Exceptional Events Rule (40 C.F.R. § 50.14) to high wind dust events and not for wildfire smoke events.

OLSAP used numerous resources, including published literature, regulatory documents, presentations, and public comments in its information gathering. The panel held several information-gathering meetings, including a 2-day open session at Owens Lake May 29–30, 2024, where several science and policy experts, Tribal representatives, and environmental organizations presented to the panel. Another open information gathering session occurred in Los Angeles on July 30, which included presentations from experts on the hydrology, climatology, and emissions of Owens Valley. Another open information gathering session occurred on October 7 and focused on exceptional events in Imperial County, Clark County, and the South Coast Air Quality Management District. Virtual open

---

[2] The focus on 2–3 km from the regulatory shoreline follows the 2016 State Implementation Plan (GBUAPCD 2016) that was based on an analysis by the Maricopa Association of Governments, which found source impacts decay by about a factor of "10 between 0 and 500 meters, between 500 and 2,800 meters, and between 2,800 and 30,000 meters." This analysis is likely valid for smaller source areas, but may not be for more disperse sources, or when those emissions are channeled down the valley.

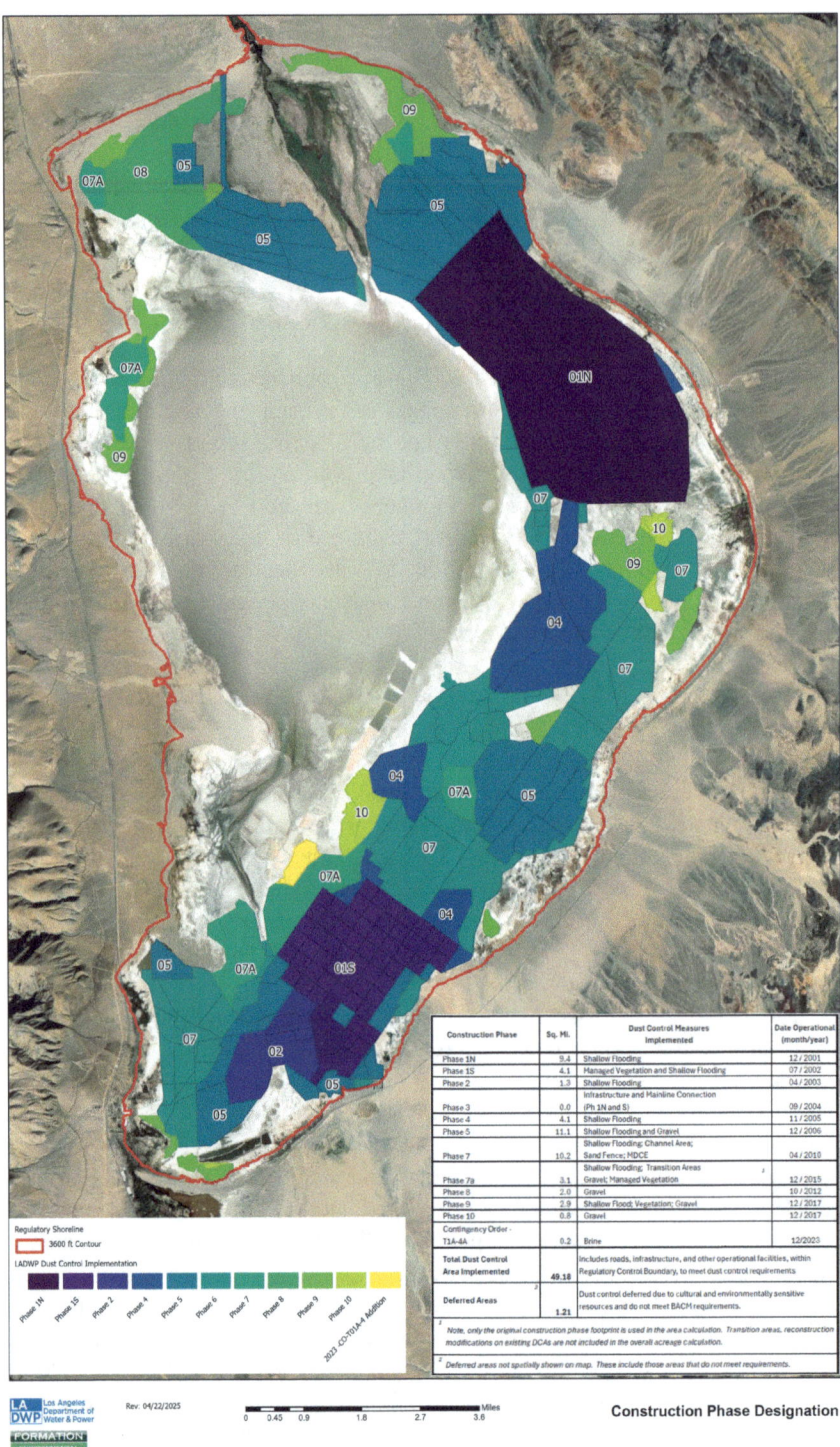

FIGURE 1-2 Dust control measures construction phase designations on on-lake sources.
NOTES: Only the original dust control phases are included and several of the dust control areas have been transitioned to other types of dust control or have been rebuilt/redesigned. Changes to acreages caused by redesigns are not shown or included in the acreage calculation.
SOURCE: Arrash Agahi, LADWP, personal communication, January 2025.

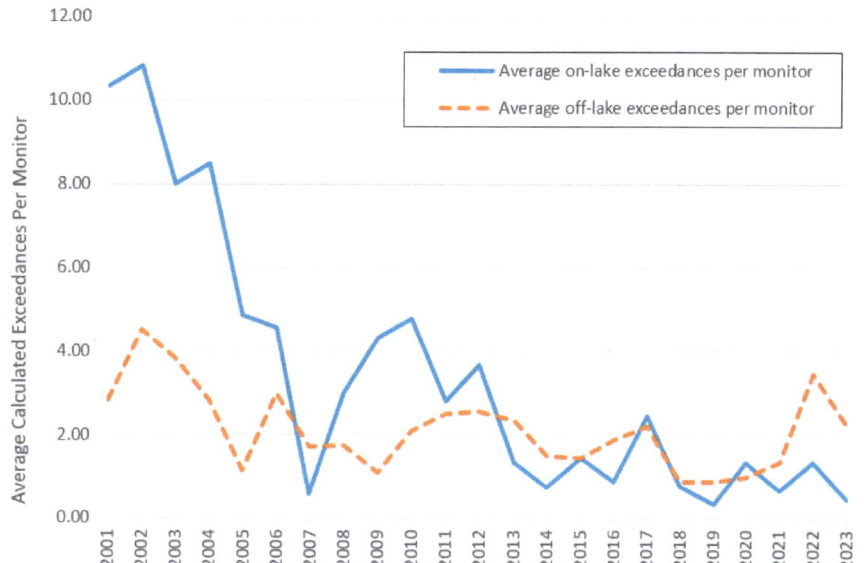

FIGURE 1-3 Calculated annual on-lake and off-lake exceedances for all monitoring sites based on consistent metrics over time divided by the number of active monitors for a given year.
NOTES: These data do not reflect official exceedance counts, but this analysis was conducted to understand long-term trends using consistent metrics. This plot takes into account the fact that monitors have been added over time, and some monitors were temporarily removed. For example, the Lizard Tail and North Beach monitors were not installed until 2008–2009. The Flat Rock PM_{10} monitor, which was in place from 2001 to 2011, was replaced by the Mill monitor in 2010. The Mill and North Beach monitors were temporarily removed from December 2012 through August 2014, and the Dirty Socks monitor was removed from December 2012 through December 2014. Exceedance counts and on- and off-lake attribution are based on screened wind direction and a threshold of 150 $\mu g/m^3$ for exceedances. For all years, on- and off-lake exceedance counts were purely classified by screened wind directions and may have included wildfire smoke events and regional events. For all hours without data when the wind was coming from an opposite direction (on- or off-lake), a background concentration of 20 $\mu g/m^3$ was assumed. Many years have incomplete data and are not a true indication of annual exceedance counts, especially in years in which a monitor was installed or removed. A statistically significant decreasing trend in the number of exceedances per monitor was determined for on-lake emissions (p-value<0.01) but the trend for off-lake exceedances was not as significant (p-value=0.03) using the non-parametric Mann-Kendall statistical test. However, if the first 2 years of data (2001–2002) are removed, the off-lake trend is no longer significant (p=0.15).
SOURCE: Generated by the panel. Data from Chris Howard, GBUAPCD, personal communication, March 2024.

sessions were held on July 12, July 26, September 6, and December 16, 2024, to investigate air quality data, strategies for engaging with Tribal communities, and dispersion modeling results in more detail.

In a few cases, the panel completed original data analyses. The results of these analyses were replicated by fellow panel members to verify accuracy. This involved multiple panel members and project staff replicating the classification of data in the District's exceedance database, and the panel also performed a comparative analysis of emission fluxes, including derivation of equations and code, shown in Appendix A.

OUTLINE OF REPORT

The report is organized into six chapters. Chapter 2 discusses the system-level context for off-lake PM_{10} sources. Chapter 3 explores the sources of PM_{10} exceedances in the OVPA and provides recommendations on how to improve monitoring and modeling. Chapter 4 provides several conclusions on the origin and evolution of

> **BOX 1-1**
> **Statement of Task**
>
> The Owens Lake Scientific Advisory Panel (OLSAP) of the National Academies of Sciences, Engineering, and Medicine will evaluate off-lake sources of airborne dust in Owens Valley, California. Off-lake dust sources are those that are located at elevations above the 3,600-ft regulatory shoreline of Owens Lake. OLSAP will focus on those sources for which available data indicate they are potentially significant contributors to high PM_{10} concentrations in the Owens Valley Planning Area (OVPA). (PM_{10} refers to airborne particulate matter with an aerodynamic diameter of 10 micrometers or smaller.)
>
> As part of its assessment, and to the extent sufficient data are available, OLSAP will summarize the overall nature and character of off-lake sources. The majority of the local off-lake sources fall within 2–3 km from the shoreline as used in the air modeling conducted for the 2016 State Implementation Plan (SIP) for achieving and maintaining the National Ambient Air Quality Standards (NAAQS) in the OVPA. In addition, the panel will address the following aspects with respect to off-lake dust sources:
>
> - Develop a summary of previous research, studies, and available information on off-lake dust sources in the OVPA that synthesizes the overall character of off-lake dust sources, including (but not limited to):
>   A. Physical setting and other characteristics, such as geomorphological features.
>   B. Extent and distribution of off-lake sources.
>   C. PM_{10} impacts of off-lake sources, with impacts defined as causing or significantly contributing to exceedances of the PM_{10} NAAQS in the OVPA.
>   D. Past and expected future changes in off-lake dust sources over time. Consideration of changes may include emissivity, number, extent, and distribution of off-lake sources.
> - Describe the origin of regulatorily significant off-lake dust sources and how they developed over time.
> - Describe potential options and recommendations to better characterize, monitor, and track changes in the PM_{10} contributions of off-lake sources to exceedances of the PM_{10} NAAQS in the OVPA.
> - Describe possible dust control measures that could be applied to significant off-lake sources, with consideration of feasibility.
> - If appropriate, discuss the applicability of U.S. Environmental Protection Agency's Exceptional Events Policy for excluding air quality monitoring data affected by dust emissions from off-lake sources from use in determinations of whether the OVPA was in compliance with the NAAQS.

off-lake sources and recommendations for how to advance this research. Chapter 5 discusses the applicability of the Exceptional Events Rule to exceedances from off-lake sources, and Chapter 6 is a solution-focused exploration of potential dust control options that could be applied, if deemed necessary.

TABLE 1-1 The Off-Lake Source Categories Relevant to this Task

PM_{10} Source Categories			Details	Comments	Relationship to Task
Off-Lake	Local (within the OVPA)	Local off-lake sources	Sources of dust emissions above 3,600 ft, adjacent or near Owens Lake, primarily located within 2–3 km of the shoreline.	The District has identified that areas adjacent to Owens Lake are causing and/or significantly contributing to PM_{10} exceedances. These sources include local dunes, sand sheets, flash flood deposits, and other open areas.	These **are** the focus of the District's request to OLSAP.
		Permitted stationary sources	Local sources of air pollution permitted by the District (e.g., mines, engines, landfills).	District has permits and enforcement for these sources.	These are not the focus of the District's request to the OLSAP. However, the emissions of these sources may be relevant to the discussion.
		"Other" local sources	This a large group of sources grouped together here for efficiency. Woodburning devices (fireplaces and stoves), transportation, unpaved roads, paved road dust, tailpipe emissions, prescribed and agricultural burning.	These are local sources that exist but have not been identified by the District as causing or significantly contributing to PM_{10} exceedances in the OVPA.	These are not the focus of the District's request, but discussion of these events may be relevant to the discussion to establish they are not contributing significantly to PM_{10} emissions.
	Regional (outside the OVPA)	Regional dust events	Dust events from outside the OVPA.	These events are typically mixed-source events, with additional contributions from local sources.	These are not the primary focus of the District's request, but discussion of these events, the sources, and the PM_{10} contributions to monitors in the OVPA are relevant to the discussion.
		Wildfire smoke events	Smoke from wildfire events.	Typically regional but could be caused from a local wildfire within the OVPA. Also, as has happened in rare occasions, event could be on-lake if it occurs below 3,600 ft.	These are not the focus of the District's request to the OLSAP.

NOTES: Some PM_{10} exceedances are caused by primarily one of the above categories; others may be caused by any number of combinations of the above sources.
SOURCE: Phillip Kidoo, GBUAPCD, personal communication, May 9, 2024.

2

A Systems-Level Understanding for Off-Lake PM$_{10}$ Sources

Effective management of dust emissions is supported by a systems-level understanding of the many factors involved in particulate matter emissions. The distribution of potentially emissive particles and their movement depend on multiple processes (e.g., geology, climate, hydrology, ecology) that vary across spatial and temporal scales, and the interaction of multiple landscape features (e.g., rivers, lakes, playas, alluvial fans, sand dunes). This chapter discusses the factors and processes influencing PM$_{10}$ emissions in the Owens Valley Planning Area (OVPA) followed by important context associated with key elements of the system, including geology, climatology, meteorology, hydrology, vegetation, and communities.

OVERVIEW OF CONTROLS OVER OWENS VALLEY PM$_{10}$ EMISSIONS

In the OVPA, multiple factors affect PM$_{10}$ emissions over space and time, including geology, climatology, meteorology, hydrology, and vegetation (Figures 2-1, 2-2, and 2-3). Humans are also an important part of this landscape because Tribal and other local communities depend on the local ecosystem for food and water, along with recreational, economic, and cultural resources; as such, these anthropogenic influences can affect PM$_{10}$ emissions. The valley also has a long history of pre-colonial Tribal water distribution and ecosystem stewardship in addition to water diversions from the Owens Valley over the last century.

Broadly, Owens Valley reflects millions of years of evolution in the geological, climatological, hydrological, geomorphic, and ecological sub-systems in the region, with interconnected and often interdependent landscapes and features. For these reasons, land use and natural resource managers continue to face steep challenges when tasked with restoring landforms or ecological sub-components within broader, dynamic landscapes, particularly in the context of climate change.

GEOLOGICAL CONTEXT

The landscape of the Owens Valley is a basin-and-range valley similar in some ways to other settings in the southwestern United States with broad valleys bounded by near parallel, north-south oriented mountain ranges. This type of landscape is the product of crustal extension over millions of years to produce faults and blocks of the Earth's crust that shift in orientation (tilting) and relief (elevation) relative to one another over relatively short distances (Eaton 1982). The Owens Valley is a fault-bounded depression, or graben, flanked by uplifted and tilted

A SYSTEMS-LEVEL UNDERSTANDING FOR OFF-LAKE PM$_{10}$ SOURCES

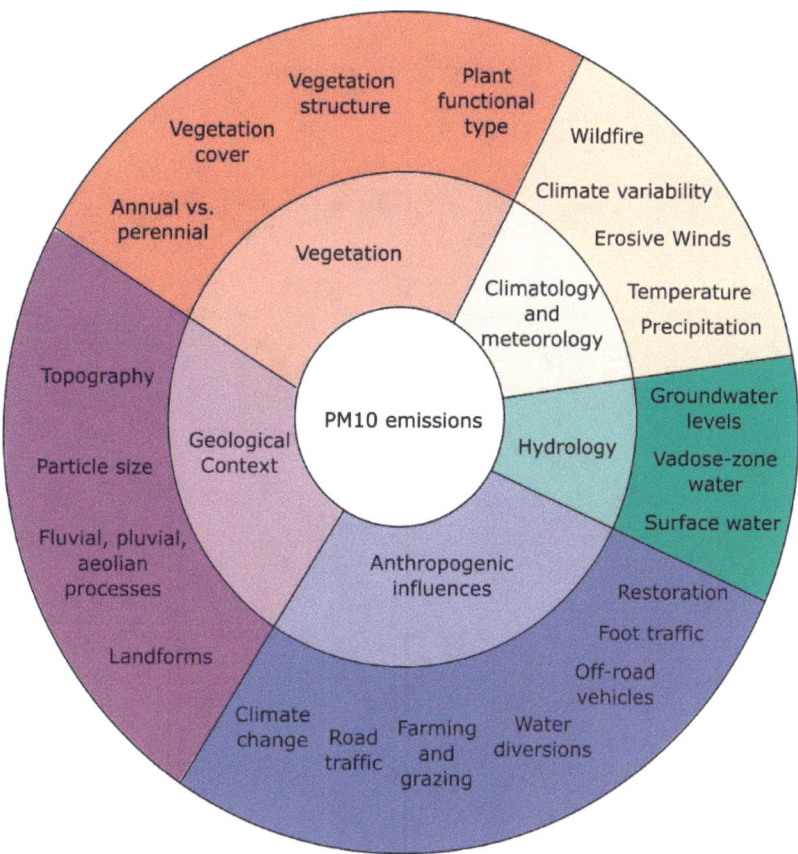

FIGURE 2-1 Processes and controls that influence PM$_{10}$ emissions in the OVPA.

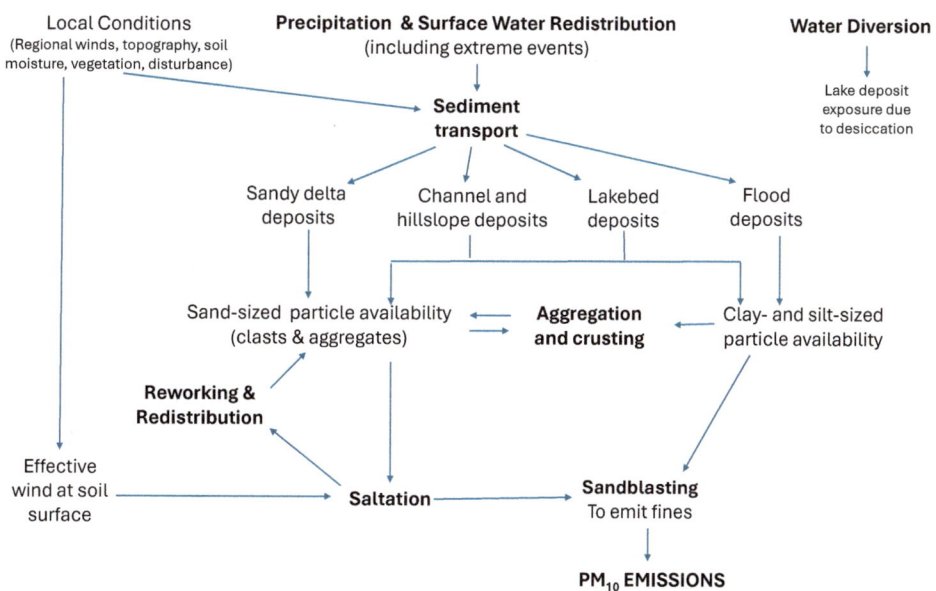

FIGURE 2-2 Controls and processes (in bold) leading to PM$_{10}$ emissions in the Owens Valley.

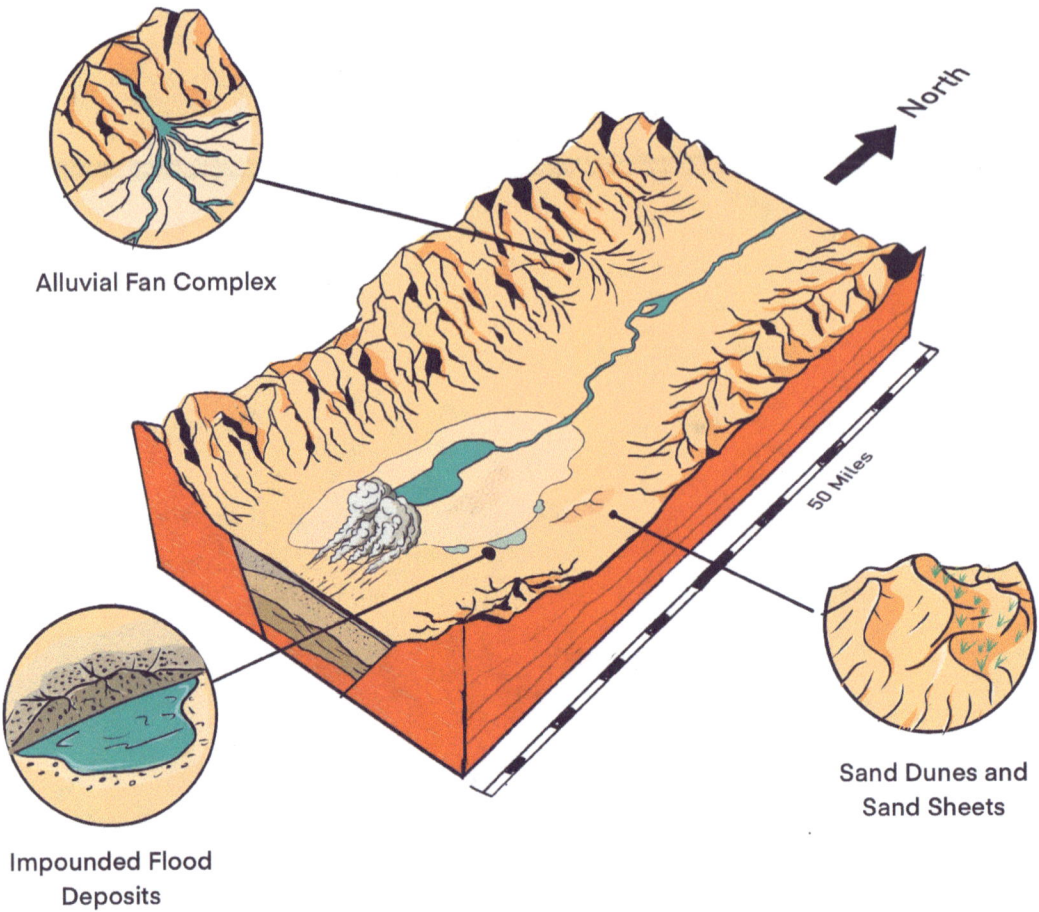

FIGURE 2-3 Conceptual model of the OVPA showing major landscape features and sources of dust emission (e.g., flood deposits impounded behind natural ridges, sand dunes, sand sheets, and alluvial fan complexes).[a]

[a] This figure was updated after release of the report to make corrections to geologic features.

crust of the southcentral Sierra Nevada Mountains to the west, the White-Inyo Range to the east, and the Coso Range to the south and east.

In terms of overall relief, the Owens Valley is among the deepest valleys in the United States, with an elevation difference of roughly 10,500 ft (3,200 m) between the valley floor and Mount Whitney, the tallest peak on the west side, which stands at 14,500 ft (4,420 m). Thus, drainage basins that connect water and sediment flows from the mountains to the valley floor are typically very energetic and have been for millions of years. Even prior to water diversion, Owens Lake had been relatively shallow due to high rates of sedimentation from erosion of nearby mountain ranges and sediment delivery from the Owens River. Because of its shallow nature, the effects of climatological changes are amplified at the lake.

For about 2.6 million years, Earth's climate has been oscillating between warm and dry interglacial periods and cooler and wetter glacial periods due to changes in Earth's orbit and axis (Milankovitch 1941). The Sierra Nevada was largely glaciated during the cooler and wetter periods (Basagic and Fountain 2011), which led to increased runoff and high water levels that would periodically overflow into neighboring chains of lakes (Bacon et al. 2020; Gale 1914). Overflow water from the paleo-Owens Lake likely stopped 11,000 to 12,000 years ago (Rosenthal et al. 2017), when it became a terminal lake (Bacon et al. 2006, 2018; Benson et al. 2002; Li et al. 2000; Smith, Bischoff, and Bradbury 1997; Smoot et al. 2000). Between 1872 and 1878, Owens Lake was a closed basin lake spanning more than 108 square miles (280 km²) with a water depth of 48.9 ft (14.9 m) (Gale 1914; Lee 1915). Major water diversions began in 1913 after the construction of the Los Angeles Aqueduct system. By 1931, Owens Lake had largely evaporated, leaving a remnant brine pool and the playa environment known today (Smith, Bischoff, and Bradbury 1997). The result of this geologic history is a complex and continually evolving spatial mosaic of different landforms in the valley fill deposits that were produced and are reworked by slope (colluvial), riverine (fluvial and alluvial), wind (aeolian), and glacial actions. In one location there may exist layers of deposits that vary in particle size, composition, and depositional structure based on differences in their origin. Thus, the Owens Valley landscape is composed of a mosaic of different surface types with differing potential to produce dust.

Dust Emission Processes

Dust emission occurs when the shear force exerted on a particle exceeds the forces attaching the particle to the surface (such as gravitational or electrostatic forces). Sand-sized particles from 63 to 100 micrometers have the lowest threshold shear speeds, and therefore, if they are present, these "efficient saltators" are the first to move, undergoing "saltation," or movement in arc-shaped hops within a few centimeters of the surface (Figure 2-4) (Bagnold 1941; Gillette et al. 1997). These particles can be single clasts or aggregates of smaller particles. Saltators ejected from the surface follow ballistic trajectories before falling back to the surface, where they strike other particles and/or disaggregate into finer particles. The impacted particles, if similarly sized or larger, can "creep" or roll along the surface, whereas smaller particles (including PM_{10}) may be ejected either from within the surface

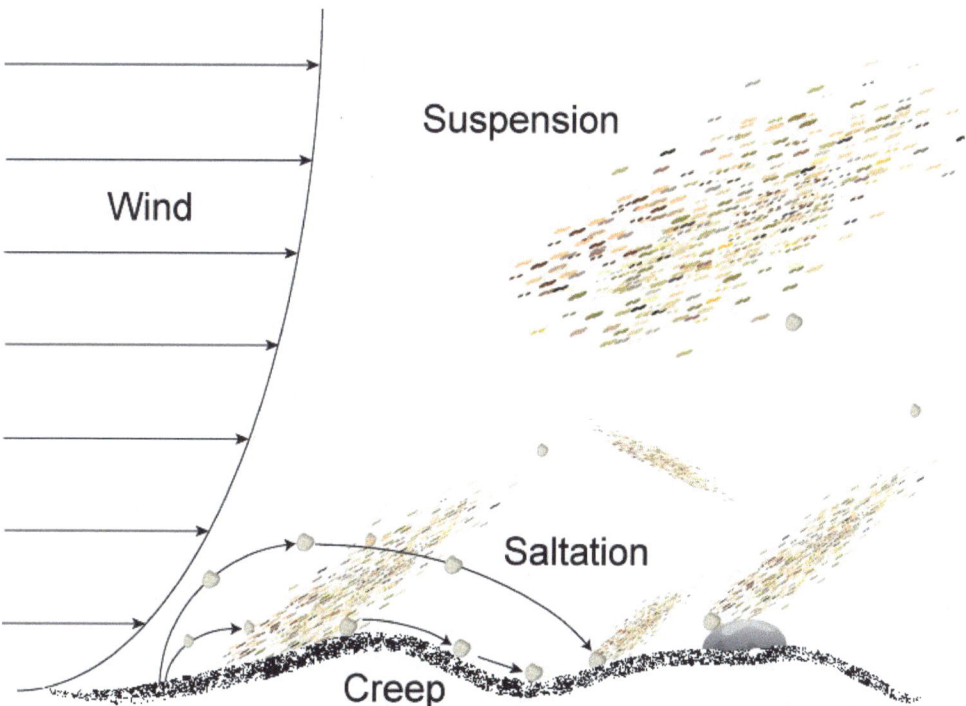

FIGURE 2-4 Wind-induced particle movements.
NOTES: The arrows on the left represent the logarithmic decrease of wind speed toward the bed surface. Larger particles move in creep mode; fine and medium sand-sized particles move by saltation mode, and smaller particles enter suspension where they are transported away. Image is not to scale.
SOURCE: Zobeck and Van Pelt (2011).

sediment matrix or as fragments from colliding particles. These finer particles can be entrained into the turbulent air stream in "suspension" (i.e., "dust" or "emitted PM_{10}"; Figure 2-4; Bagnold 1941). Thus, if particles capable of saltation and emission as PM_{10} are present, dust emission can occur on any surface, including natural surfaces around the lake. The amount of dust emission flux (F_d) from a surface is proportionate to the saltation flux (Q), both of which are a function of the wind shear velocity (u_*) at the surface. Generally, the volume of fine dust particles emitted from a surface by saltation takes the form, $F_d \propto u_*^i$, where i = 3 for harder crusted surfaces and i = 4 for soft, loose sand or soil surfaces (Lu and Shao 1999; Sweeney 2022). Dust emission is a natural process in semiarid landscapes with competent winds and sufficient particles for saltation and suspension. Although direct aerodynamic entrainment of dust-sized particles without saltating sand is possible, it is believed to be relatively rare on natural surfaces with varying roughness elements (e.g., plants, mixed sediments, dunes) and soil moisture conditions (Sweeney 2022).

Potentially Emissive Landforms

Surface water has the potential to move both fine dust-sized particles and larger sand- and gravel-sized particles. This is particularly true in overland and channelized floodwaters moving from the mountain front down toward the bottom of the valley. During higher energy flood events, particles of all sizes (including boulders) can be transported by water, with larger particles deposited on the upper alluvial fans and finer particles moved to more distal reaches of alluvial fans and onto the valley floor. Thus, the distal edges of alluvial fans tend to be finer grained than areas closer to the mountain fronts, and fans generally tend to be composed of clasts too large to be moved by wind. In lower energy situations, such as during small flow events or during the waning stages of higher magnitude floods when the flow slows, finer particles carried by the water will settle out. This can happen in the channel or in localized depressions behind natural or anthropogenic impoundments where water pools and deposits fine material capable of producing dust once the area dries. Thus, although alluvial fans are typically made of clasts too large for wind to move, channels inset in fans and distal areas of fans often contain a mix of particle sizes, including those moveable by wind through saltation and suspension, creating areas within the fan that may experience considerable aeolian transport and PM_{10} emission.

Fine, dust-sized particles (i.e., clay and silt) may also be transported to lower energy environments, like standing water and lakes. These fine-grained lacustrine sediments can be quite cohesive, with little ability to produce dust without a disturbance to break up the soil crust. This is particularly true if binding agents (such as salts or clays) or biological crusts are present (see review and citations in Sweeney 2022). In cases where a mixture of dust- and sand-sized particles were deposited (e.g., where channelized flow empties into the lake), removal (i.e., winnowing) of fines by aeolian transport can produce a coarse sandy lag deposit with reduced ability to produce dust. This appears to be the case throughout much of the valley bottom, although areas still exist that are clearly capable of producing dust under strong winds. Similarly, most of the beach deposits found around the paleo-Lake Owens would likely have been comprised of sand- to cobble-sized clasts. However, in the millennia since the deposition of these beach deposits, the strong winds of the Owens Valley have largely removed sand-sized particles, leaving behind larger clasts immovable by wind. Nonetheless, beach ridges can produce localized depressions landward of the shoreline that are well suited as depositional basins for impounded floodwaters and other processes that deposit dust-producing sediments.

Sand dunes are typically produced from aeolian reworking of sand deposits that were initially delivered by fluvial or lacustrine processes. In the case of the Owens Valley, this process of dune formation can include sand delivery to the lakebed and channel floodplains by the Owens River and in-lake, wave-driven littoral currents that create sandy beaches, that are then re-mobilized by wind and re-deposited into dune forms downwind. Through saltation (Figure 2-4), winds competent to transport sands are also capable of mobilizing and producing finer silt- and clay-sized particles, which become deposited within dunes and associated vegetation communities. Thus, dune ecosystems are naturally capable of both storing and providing fines capable of emission as dust (Bullard and Baddock 2019; Sweeney 2022; Sweeney, Lacey, and Forman 2023). Depending on the wind regime, sand supply, and vegetation cover and type, aeolian sands can be deposited as unanchored "free" dunes (e.g., barchans, transverse dunes) or as "anchored" dunes associated with shrubby vegetation, such as parabolic dunes (Goudie 2011) or *nebkhas* (also called "coppice dunes"; Hesp and Smyth 2019; Lancaster and Bacon 2012). Vegetated dunes are typically poor dust producers because the vegetation reduces the mobility of saltators by reducing wind shear at the surface. However, vegetated dunes also serve as sinks for dust-sized particles over time either through direct capture of suspended fines by plant canopies, atmospheric (wet or dry) deposition, fluvial or pluvial deposition of fines within interdune depressions, or in situ production of fines (Bullard and Baddock 2019; Sweeney 2022). Saltating particles can sandblast vegetation, damaging or even killing it, and the loss of fines associated with dust emission can reduce important soil resources including cation exchange capacity, water holding capacity, soil nutrients, and plant seeds (Okin, Gillette, and Herrick 2006). These impacts can destabilize vegetated dunes causing re-mobilization of the underlying sand dunes, which not only liberates saltators but also exposes dust-sized particles in the newly mobilized dune field to sandblasting and emission. In the case where large amounts of fines were deposited within the dune field while it was stabilized, the dunes could become significant sources of dust indefinitely.

Sandy surface soils are not always formed into sand dunes by wind; they can also form extensive sand sheets. Active sand sheets differ from sandy soils mainly by their limited organic content and presence of ripples, which reflect frequent transport of sand grains by wind. Sand sheets tend to form not only where there is a significant source of sand, but also in areas where either wind power or sand supply is too low to create dunes. In addition, Kocurek and Nielson (1986) indicate other local factors that are often associated with sand-sheet development, including: a high water table, surface cementation or binding, periodic flooding, significant coarse-grained sandy sediment supply, and vegetation cover. For example, the sand sheet above Keeler Dunes that is moving up the Slate Canyon fan appears to exist under conditions of relatively low sand supply (mostly coarser-grained sands), periodic flooding, and sparse vegetation cover. This results in a thin layer of sand over the modern, coarser (gravel-cobble dominated) alluvial fan. Alternatively, a sand sheet might be a transient landform caused by recent exposure of considerable sand without sufficient time for transport into an organized dune field and/or other constraints that would limit the development of dunes. The northern sand sheet on the Owens Lake bed northwest of the Keeler Dunes appears to be such a landform. This area, rich in sand, forms a part of the Owens River delta in Owens Lake and the morphology of sand deposits here reflect conditions imposed by a high water table in the valley bottom, periodic flooding, availability of coarser sands (alluvial deposits of the Owens River), and the presence of binding salts. The relatively recent exposure of this deposit following diversion of water from the lake provided a sudden increase in the abundance of sand available for transport prior to dust control efforts.

CLIMATOLOGY AND HYDROLOGY

High winds and precipitation affect dust emissions in the OVPA. In the Owens Valley, high winds are frequently aligned with the axis of Owens Valley and are either from south-southeasterly or north-northwesterly directions (Zhong et al. 2008). About 40 percent of wind events over 7 m/s (16 mph) occur in the spring (March–May), about 24 percent in fall (September–November), and less than 20 percent in winter (December–February) or summer (June–August; Zhong et al. 2008). Extreme high wind events over 18 m/s (40 mph) tend to occur only a few times a year (Zhong et al. 2008). Cross-valley westerly high wind events occur less frequently but are the strongest wind events with the longest duration of sustained winds and are related to some strong dust events (Jiang, Liu, and Doyle 2011; Serafin, Strauss, and Grubišić 2017).

The crest of the Sierra Nevada in the Owens Valley region receives more than 30 inches/year (in/yr) of precipitation, with the majority occurring from October to April and small amounts falling during summer thunderstorms (Danskin 1998). The Sierra Nevada Mountains produce a rain shadow effect that results in the Owens Valley being substantially drier than the western side of the Sierras, so that the Inyo and White mountains only receive 7–14 in/yr of precipitation. The annual precipitation on the Owens Valley floor varies from 6 in/yr in the north end of the OVPA to approximately 4 in/yr around Owens Lake. Precipitation also varies with elevation, with higher elevations receiving substantially more rain than lower areas (Danskin 1998). Although precipitation is generally low in the Owens Valley, water flow from the Sierra Nevada due to snowmelt can be substantial (Hollett et al. 1991). Higher rainfall years can enhance vegetation cover and reduce dust generation, while drought can decrease vegetation cover and increase dust generation (Elmore et al. 2008). Precipitation patterns also can influence soil moisture, with drier surfaces being more emissive (Sehmel 1980).

Alluvial transport of sediments is frequently associated with extreme precipitation events that cause flooding, creating new emissive surfaces (Elmore et al. 2008). In the Owens Valley, such flooding can occur due to high and rapid snowmelt from the Sierra Nevada, or from infrequent, but intense precipitation from thunderstorms, atmospheric rivers, or tropical storms (Kim and Lowe 2004; Sahagun 2024). For example, the remnants of Tropical Storm Kay brought flash flooding during a brief but high-intensity event that impacted Owens Valley in September 2022. Recorded rainfall totals ranged from 0.22 to 1.06 inches in the immediate area around Owens Lake (National Centers for Environmental Information 2025). However, due to the observed spotty nature of rainfall and the sparse network of rain gauges in this area, higher amounts of rainfall than recorded at the gauges could have been possible. In August 2023 during Tropical Storm Hilary, weather stations around Owens Lake recorded 24-hour precipitation totals ranging from 2.25 to 5.39 inches (GBUAPCD 2023). To put these into perspective, the average annual rainfall at nearby Bishop Airport is 5.13 inches (National Centers for Environmental Information 2025).

Groundwater also has strong effects on dust emission potential by altering soil moisture, soil chemistry, soil fertility, and the type and extent of vegetation cover (Elmore et al. 2008; Goedhart and Pataki 2011). Groundwater is particularly important in the lower Owens Valley, where most streams are now diverted into the Los Angeles (LA) Aqueduct, and do not directly flow into the valley (Danskin 1998). Nearly all groundwater in Owens Valley is stored in the saturated valley fill because the bedrock that surrounds and underlies the valley fill is virtually impermeable (Hollett et al. 1991, 1989). Given the low rainfall on the valley floor itself, nearly all the recharge to the aquifer system is from infiltration of runoff from snowmelt and rainfall on the Sierra Nevada. Groundwater flows from the margins of the valley toward the center and then to the south from the Bishop basin (outside the OVPA) to the Owens Lake basin (Hollett et al. 1991).

Groundwater use in Owens Valley to the north of Owens Lake has changed over the decades (Danskin 1998). Prior to 1913, groundwater tables in Owens Valley were generally high and supported dense vegetation, and canals were used to decrease water levels in some areas to allow for agriculture. Between 1913 and 1969, the extensive surface water export through the LA Aqueduct led to severe water limitations in Owens Valley, causing many farms and ranches to cease operations, but groundwater was only occasionally pumped. In the early 1970s, groundwater levels and the acreage of native vegetation it supported were estimated to be similar to 1912–1921 values (Griepentrog and Groeneveld 1981). The opening of the second LA Aqueduct in 1970 increased export capacity from the Owens Valley by 60 percent and led to a 5-fold increase in groundwater pumping from the valley floor (Danskin 1998; Elmore, Mustard, and Manning 2003), resulting in a 35 percent reduction in native plant evapotranspiration and near elimination of some spring discharges. Large, more widespread changes in depth to groundwater were seen during the droughts of the 1980s and 1990s (LADWP and County of Inyo 1990a) with groundwater extraction peaking in 1987 and 1988 at approximately 240,000 m^3/yr (or 190 acre-feet per year; Elmore, Mustard, and Manning 2003). Decreases in spring flow and evapotranspiration served to buffer declining groundwater levels along the Owens Valley floor, but larger groundwater table declines occurred along the western alluvial fans and near well fields (Danskin 1998). From 1986 to 1992, 41 of the 171 piezometers in Owens Valley from Lone Pine to Bishop showed a decline in the water table from 3.3 to 6.6 ft (1 to 2 m); 31 wells declined from 6.6 to 13.1 ft (2 to 4 m); 12 wells declined >13.1 ft (4 m), with the greatest decreases in groundwater levels occurring near the areas of greatest pumping (Figure 2-5; Elmore, Mustard, and Manning 2003).

In 1990, a groundwater management plan—the Long-term Water Agreement—was implemented following a series of disputes surrounding alleged environmental harm from water diversion from the valley (LADWP and County of Inyo 1990a). Since this agreement was reached, groundwater withdrawals have been managed to minimize vegetation impacts, and groundwater levels have fluctuated without a long-term directional trend (Figure 2-6), although modeling by Danskin (1998) suggests that full recovery to pre-1970 groundwater levels in the OVPA north of Owens Lake has not been reached. During the 2012–2016 drought the Los Angeles Department of Water and Power (LADWP) reduced groundwater pumping below that permitted by the Long-term Water Agreement, and groundwater levels did not drop as steeply as during the 1986–1992 drought. Groundwater levels were able to recover after a few years in most wells (Owens Valley Groundwater Authority 2021). In 2022, 46 percent of groundwater extraction was used for irrigated agriculture in Owens Valley; 33

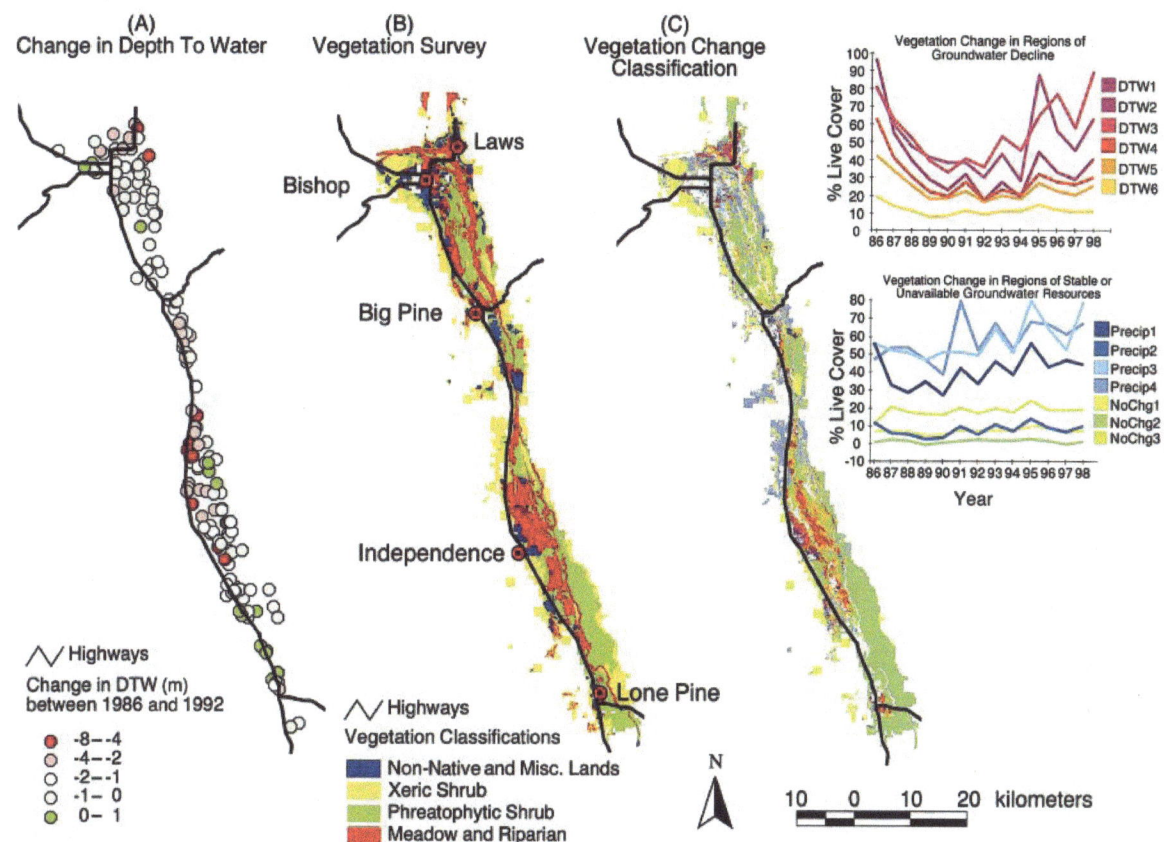

FIGURE 2-5 A) Changes in depth to groundwater (DTW) for wells in the Owens Valley; B) vegetation classifications from a vegetation survey dated 1984–1987, by which time considerable pumping had already occurred; C) 13 vegetation response classes from a remote sensing analysis in 1998 with red indicating areas with the largest change due to groundwater level decline (see details in the graphs in the upper right).
NOTES: The graphs in the upper right are the annual average percentage of live cover for each change classification in C). DTW, Precip, and NoChg designate change classes representing depth to water dependent changes, precipitation dependent changes, and static vegetation conditions, respectively.
SOURCE: Elmore, Mustard, and Manning (2003).

percent for managed vegetation and wetlands; 14 percent for municipal and domestic uses; and 7 percent for Tribal use and unspecified LADWP uses (Owens Valley Groundwater Authority 2024).[1]

In contrast to the changes up the valley, the vicinity of the lake (also known as the Owens Lake Management Area under the Sustainable Water Management Act) has seen lower levels of groundwater pumping (Owens Valley Groundwater Authority 2021).[2] Perennial and seasonal springs and seeps continue to ring the regulatory shoreline of Owens Lake (Figure 2-7), with measured water table depths varying seasonally by 0 to 5 ft (0 to 1.5 m; LADWP 2012; Victor Harris personal communication, July 30, 2024) and by as much as 15 ft across years (Figure 2-8). This and other evidence (Meyers et al. 2021) indicate generally stable groundwater levels near the Owens Lake regulatory shoreline despite the diversion of the Owens River and the groundwater pumping occurring up-valley.

---

[1] This paragraph was edited after release of the report to correct information about groundwater pumping.
[2] This sentence was edited after release of the report to clarify the geographic area described.

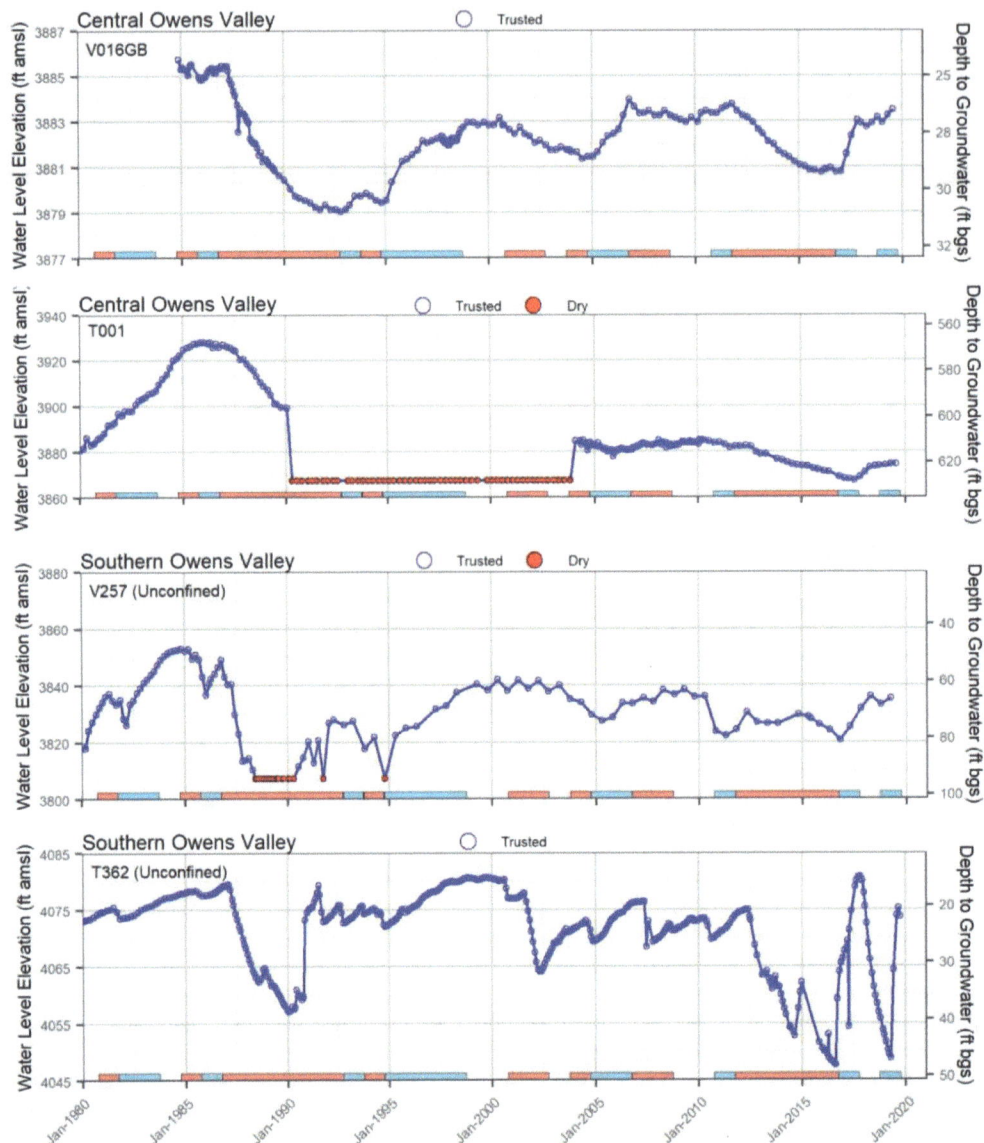

FIGURE 2-6 Groundwater levels for wells north of the lake that have long-term records.
NOTES: V016GB is near Fish Springs; T001 is near Independence; V257 is south of Independence; T362 is north of Lone Pine. See map in OVGA 2021. "Trusted" measurements refer to measured water table elevations considered reliable and accurate, in contrast to elevations that were measured when the monitoring well was dry. Most locations show fluctuating groundwater levels without a long-term directional trend. As noted by the text, major increases in groundwater pumping from the valley floor occurred in 1970, prior to these data. At the bottom of the graphs, orange bars denote dry years, and blue denote wet years.
SOURCE: Owens Valley Groundwater Authority (2021).

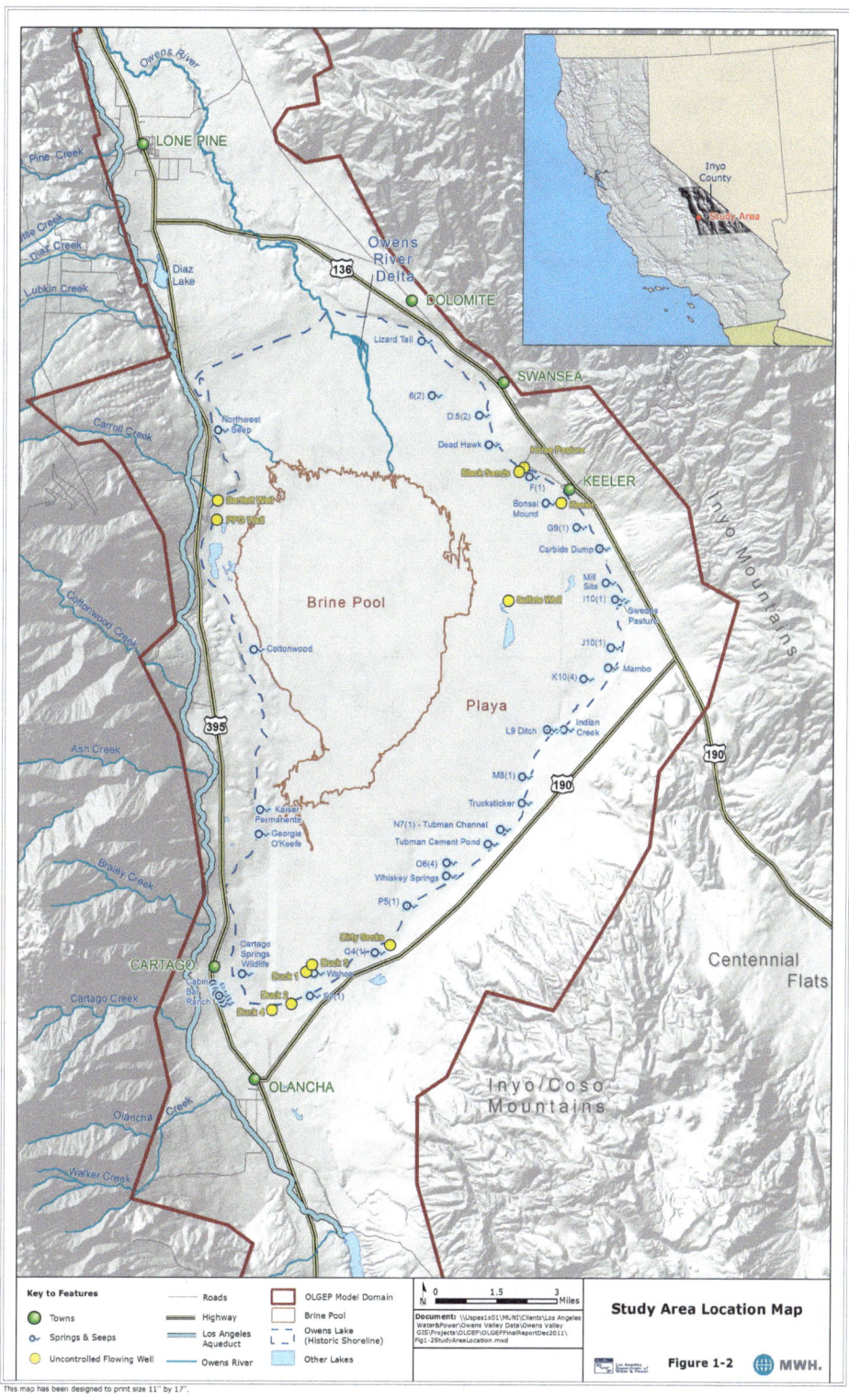

FIGURE 2-7 Wetlands, springs, and seeps that ring Owens Lake.
SOURCE: LADWP (2012).

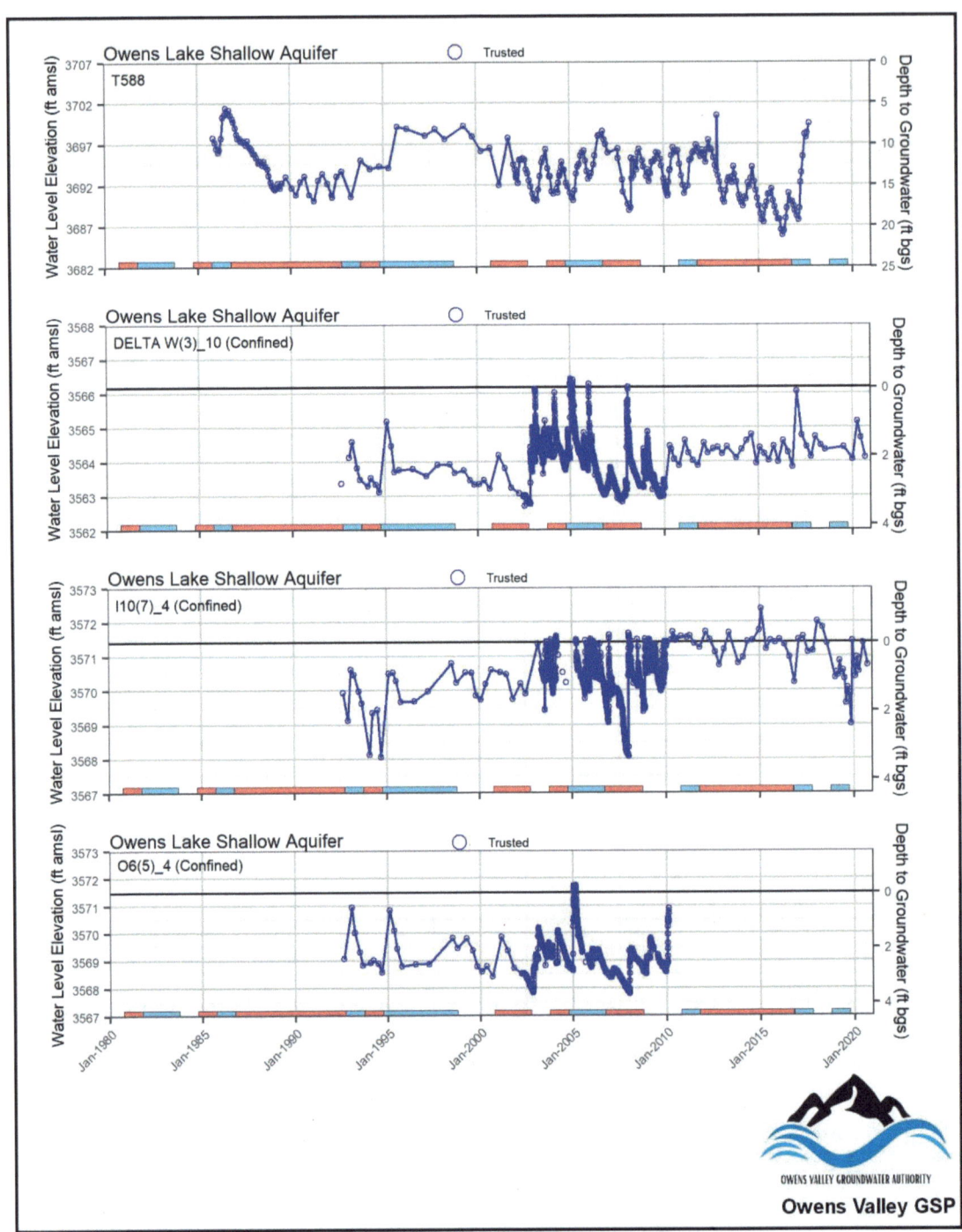

FIGURE 2-8 Trends in groundwater levels at wells near Owens Lake.
NOTES: Most locations show fluctuating groundwater levels without a long-term directional trend. The top site represents water levels for a single well screen from 30 to 40 ft below the ground surface, and the others denote data from shallow piezometers screening between 3 to 10 feet below the ground surface. T588 and DELTA W(3)10 are north of the lake; I10(7)4 is east of the lake; and O6(5)_4 is south of the lake (see map in Owens Valley Groundwater Authority 2021). At the bottom of the graphs, orange bars denote dry years, and blue denote wet years.
SOURCE: Owens Valley Groundwater Authority (2021).

VEGETATION DYNAMICS

Vegetation impacts dust emission and deposition. For example, vegetation increases the surface roughness, thereby reducing wind shear speed at the soil surface, which in turn reduces dust emission (Okin, Gillette, and Herrick 2006). Vegetation can also entrap dust and sediment from the air (Raupach et al. 2001). If changing conditions lead to lower vegetation cover, previously vegetated soils have the potential to be considerable sources of airborne dust, and decomposing organic matter can form dust-sized particles that are relatively easy to suspend in the airflow due to their low density. Additionally, the incursion of invasive species can impact the local ecology and in some cases may lead to reduced vegetation cover or increased disturbance to the ground. For example, Russian thistle (*Salsola* sp.) is reported to be commonly found in ditches, roadsides, and cattle grazing areas east of Owens River. Because it is an annual plant, it does not contribute to sustained stabilization of the landscape, and concerns have been raised that tumbleweeds from this thistle disturb the ground, leading to increase dust emissions (Mel Joseph, Lone Pine Paiute Shoshone Tribe, personal communication, August 2024).

Soil salinization has the potential to control vegetation cover, but soil in Owens Valley is generally not saline enough to limit the existing vegetation, except in unique topographic situations that allowed for high deposition and low resuspension of dust (Quick and Chadwick 2011; Reheis 1997). Instead, water availability is the key driver of vegetation composition and management in the Owens Valley (LADWP and Ecosystem Sciences 2010). Access to groundwater can be especially important in Owens Valley, where evapotranspiration from the valley floor is 3- to 6-fold greater than precipitation (Hollett et al. 1991). Upland plant communities, such as dryland/xeric scrub, are dependent on precipitation because their roots lack access to deep groundwater and tend to have low percent cover (median = 6 percent) and high variability from year to year. These areas with deep groundwater are dependent on precipitation and horizontal redistribution of water by hillslope hydrologic processes, making them susceptible to drought (Elmore, Mustard, and Manning 2003). In contrast, areas with high groundwater levels near seeps and springs support wetland vegetation and areas with groundwater depths of up to 16 ft can support deep-rooted plants, termed phreatophytes, which can access groundwater directly (Cooper et al. 2006). These groundwater-supported plant communities, such as high-groundwater alkali scrub and alkali meadows, have higher and more stable plant cover (Richards et al. 2022). The varying groundwater depths in the Owens Valley give rise to different ecological communities (Figure 2-9), which impacts dust emission because groundwater-dependent vegetation has a higher percent cover than upland vegetation types (Richards et al. 2022). For example, one study showed approximately 30 percent cover in shrub communities with >6.5 ft (2 m) depth to groundwater, versus >50 percent cover in grass-dominated communities with <3.2 to 7.5 feet (1 to 2 m) depth to groundwater (Goedhart and Pataki 2011).

The most common groundwater-dependent vegetation type in the Owens Lake area is greasewood (*Sarcobatus vermiculatus),* making up 22.9 percent of the vegetation in the mapped groundwater-dependent vegetation areas (Stillwater Sciences 2021). *Suaeda nigra* (bush seepweed) and *Atriplex parryi* (Parry's saltbush) are also common in areas where the groundwater is less than 16 ft (5 m) from the surface. Saltgrass (*Distichlis spicata)* can grow with these shrubs in stable areas, and shrubs transition to saltgrass meadows where there is shallower groundwater. In the upland community (where groundwater is deeper than 16 ft [5 m]), vegetation is dominated by three saltbush species (*Atriplex confertifolia, Atriplex parryi, Atriplex hymenelytra)* and can also include creosote bush (*Larrea tridentata)* and sometimes *Sarcobatus baileyi* and *Suaeda nigra* (Richards et al. 2022).

While the type and extent of vegetation around much of Owens Lake has likely remained relatively consistent given the stable groundwater conditions, major changes in vegetation have occurred to the north of the lake in Owens Valley. These up-valley areas had been covered by "swampy lowlands" as recently as the early 1900s, (Steward 1933) and they were dominated by large alkali meadows (Benson et al. 2002), requiring canals to drain areas to allow for cultivation (Danskin 1998). By 1920, almost 40 square miles (or 10,000 hectares) of the Owens Valley were cultivated, with 80 percent of those lands abandoned by 1935 due to insufficient water with additional abandonment occurring in the decades that followed (McLendon, Naumburg, and Martin 2012). In the early 1970s, the acreage of phreatophytic plants was similar to that observed between 1912 and 1921 (Griepentrog and Groeneveld 1981). However, once extensive groundwater pumping started in the 1970s, 26,000 acres of the Owens Valley showed a 20–100 percent loss in vegetation cover by 1981 (Griepentrog and Groeneveld 1981). Changes in vegetation associated with changes in depth to groundwater by the combined impacts of precipitation deficits and

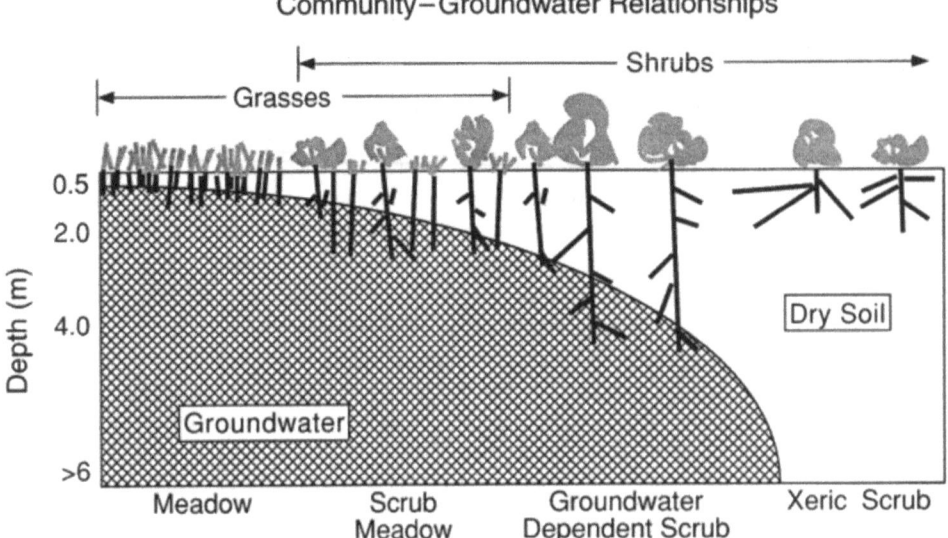

FIGURE 2-9 Plant communities in Owens Valley vary based on depth to groundwater.
SOURCE: Elmore, Mustard, and Manning (2003).

groundwater pumping were documented in the Owens Valley between 1980s and the early 2000s (Figure 2-10). These studies showed both decreased live plant cover and replacement of grasslands by shrubs in Owens Valley (Elmore, Mustard, and Manning 2003; Griepentrog and Groeneveld 1981; Manning, 1997). Groundwater-dependent meadow communities exhibited considerably greater impact due to changes in depth to groundwater than shrub communities (Figure 2-10; Elmore, Mustard, and Manning 2003). In areas adjacent to wellfields, the cover of perennial plants decreased, and grass was replaced by shrubs, while this did not occur in control areas that were not pumped (Jabis 2011; Manning 1999). At present, the dominant vegetation type in the OVPA is characterized as arid or semiarid scrub (LADWP and County of Inyo 1990a).

With the 1990 Long-term Water Agreement, LADWP agreed to manage water resources to avoid environmental impacts, such as decreases in vegetation cover and changes in vegetation type (LADWP and County of Inyo 1990a). The baseline vegetation types and cover guiding these actions were those documented between 1984 and 1987, even though substantial groundwater withdrawal had led to large changes in vegetation type and coverage by this time, and many of these baseline habitats were substantially degraded (Borgias 2024; Danskin 1998). A main tenet of the Long-term Water Agreement is that the extent of five classes of vegetation be maintained or improved to prevent conversion to more water-limited vegetation types. Even with the limits on groundwater pumping, vegetation recovery has been slow, and vegetation has converted to more water-limited types (Borgias 2024; Jabis 2011; Manning 1999). As part of the Long-term Water Agreement, LADWP is required to mitigate some areas of vegetation loss, through efforts such as planting and/or irrigation. Of the 66 required environmental mitigation projects, 9 are complete, 51 are implemented and ongoing, and 6 were implemented but did not meet goals (LADWP 2024a). Some sites have generally achieved the baseline plant cover goals corresponding to the 1984–1987 data, but a number have not, and very few were able to do so during the 2012–2016 drought (LADWP 2024a). In areas with long-term water table drops, grass regrowth after fire is minimal, leading to erosion and conversion to xeric shrubland (Pritchett and Manning 2009). Similarly, on agricultural lands abandoned around high groundwater pumping in 1970, several areas have not had any perennial vegetation recovery, and average recovery rates on drier sites are 60 percent of those of wet sites (McLendon, Naumburg, and Martin 2012).

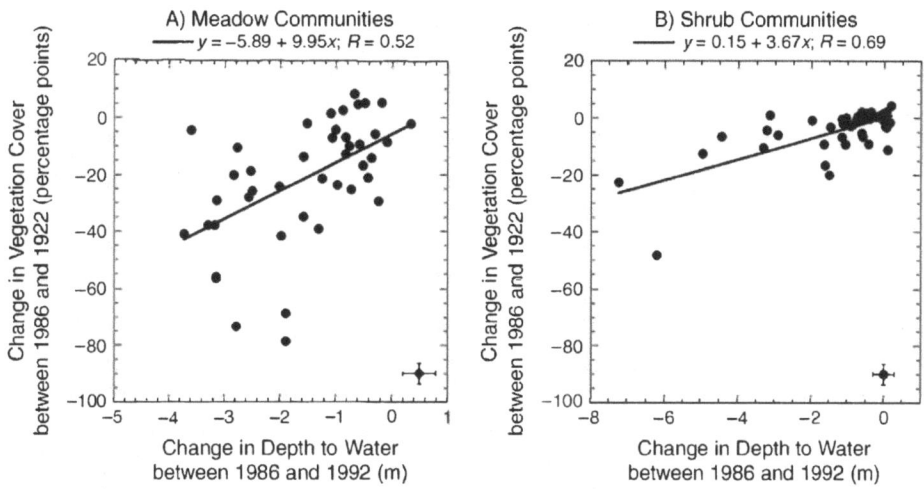

FIGURE 2-10 Landsat-derived percent live cover for meadow and shrub communities correlated with change in depth to water.
NOTES: Correlations were statistically significant (p <0.05). Error estimates are shown in the lower right of each graph. Y-axis should read "Change in vegetation cover between 1986 and 1992."
SOURCE: Elmore, Mustard, and Manning (2003).

ANTHROPOGENIC DISTURBANCE OF VEGETATION AND THE SOIL SURFACE

Anthropogenic disturbances have the potential to increase dust emissions by nearly an order of magnitude either through changes to the soil surface or vegetation (Van Pelt et al. 2020). Any activity that removes vegetation has the potential to increase dust emissions as vegetation decreases the availability of saltating particles (i.e., a supply limitation) and reduces wind shear stress on the surface (i.e., a transport limitation; Sherman and Ellis 2022). Additionally, Gillette et al. (1980) showed that disturbance of the soil surface could dramatically reduce the threshold for aeolian transport through destruction of protective physical crusts (see also Belnap et al. 2007; Khatei et al. 2024; Van Pelt et al. 2020). Baddock et al. (2011) showed that soil surface disturbances could substantially increase dust emissions, partially through reducing the threshold for transport and partially by increasing the number of particles available for emission.

Often a disturbance might impact both vegetation and soils. For instance, off-highway vehicle (OHV) activities can kill vegetation while also breaking up physical and biological crusts and grinding stable aggregates to produce dust-sized particles. OHV traffic is permitted in the Olancha Dunes Recreation area, and vehicles driven onto fine-grained, dry backwater lakes for sport have been reported to create dust emissions (Mel Joseph, Lone Pine Paiute Shoshone Tribe, personal communication, August 2024). Similarly, agriculture removes vegetation, disturbs soil crusts, removes non-erodible clasts, and breaks down soil aggregates, while grazing reduces vegetation cover through herbivory. Animal hooves are also known to break up physical and biological crusts.

CLIMATE CHANGE

Dust control may become more challenging under the changing climate conditions projected for the Eastern Sierra with projections of a 6–10 degrees Fahrenheit increase by 2100 (Dettinger et al. 2018). While annual precipitation in Owens Valley has been relatively consistent since the 1940s with no clear decadal trends (Figure 2-11), predictions for future trends in precipitation across the Sierra Nevada range from −5 percent to

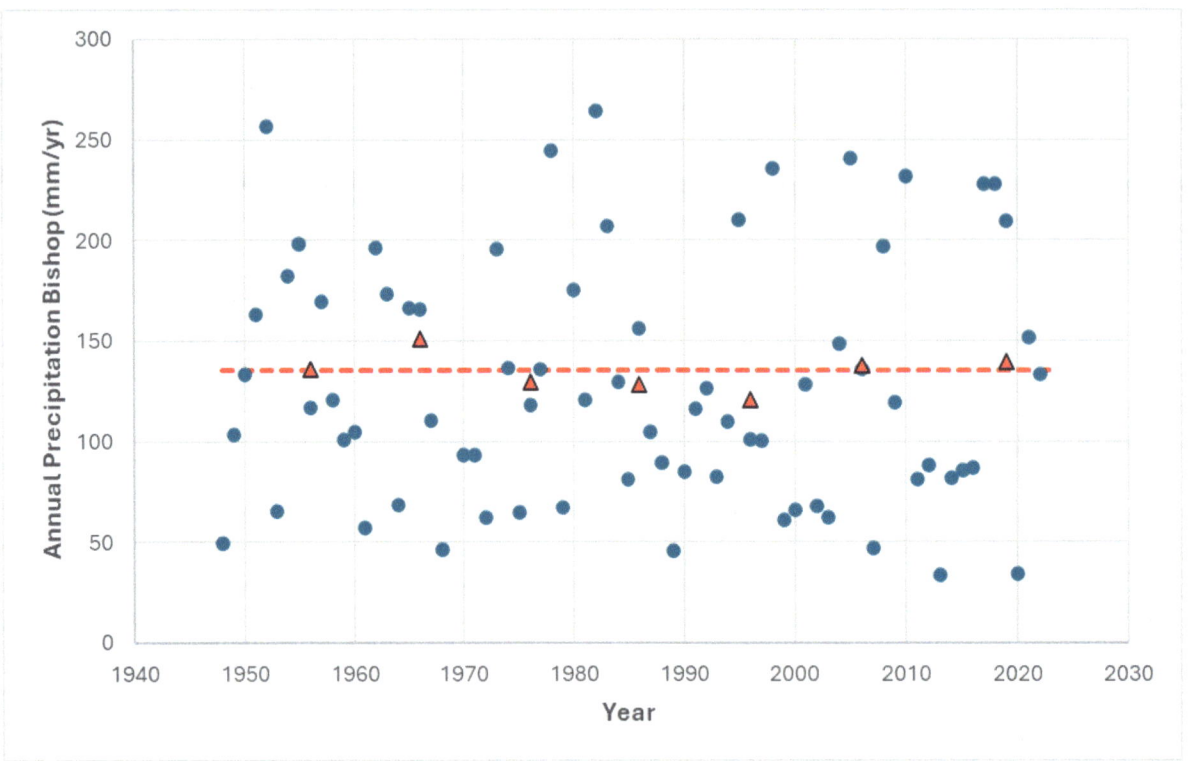

FIGURE 2-11 Annual precipitation records from Bishop Airport station with annual values (blue disk) and decadal averages (red triangle) indicating interannual variability with no persistent trend.
NOTES: Red line is the mean of the annual values.
SOURCE: Data from AgACIS at Bishop Airport.

+10 percent by 2070–2099 (Dettinger et al. 2018). Modeling by the Owens Valley Groundwater Authority (2021) predicted a 6 percent increase in precipitation along with a 19 percent increase in evapotranspiration, resulting in an average 3 percent increase in groundwater recharge by 2045. More precipitation is expected in the form of rain rather than snow, leading to earlier stream flow (Harpold et al. 2015) into the Owens Valley.

Although the average annual precipitation may not change considerably, precipitation extremes are projected to increase (Swain et al. 2018) with more precipitation falling in large events (e.g., atmospheric rivers; California DWR Climate Change Technical Advisory Group 2015; Espinoza et al. 2018; Gershunov et al. 2019; Huang, Stevenson, and Hall 2020). With a 5–30 percent increase in large storms (Dettinger et al. 2018) and a 40 percent average increase in peak hourly precipitation rates (Huang, Stevenson, and Hall 2020), frequent flood events that cause sediment movement that may then remobilize as dust are more likely (East and Sankey 2020). With climate change, the length of time between rain events is expected to increase (Polade et al. 2017), which together with increased evapotranspiration contributes to a projected 50 percent decrease in streamflow during drought years (Alex Hall, UCLA, personal communication, July 2024). Vegetation cover may decrease during extended dry periods (especially under hotter temperatures), making the surfaces more vulnerable to erosion and dust emission.

TRIBAL KNOWLEDGE AND CULTURAL PRIORITIES

The *Nüümü* (Owens Valley Paiute) named the Owens Valley "*Payahuunadü*" or "land of flowing water" (Owens Valley Indian Water Commission 2024). A date of 1000 CE is often given for the arrival of the Paiute into the Owens Valley (Varner 2009). However, human occupation has occurred over a much longer period in the southwest United States, as far back as 23,000 years ago (Pigati et al. 2024). Owens Valley was one of the most densely settled regions of the entire Great Basin and was unique in the Great Basin due to its permanent villages (Lawton et al. 1976). These settlements were supported by a mix of hunting, gathering, and agriculture, partially supported by extensive water flow from the Sierra Nevada, resulting in the prevalence of the "swampy lowlands of Owens Valley where it is obvious that moist soil—a natural irrigation—produces a very prolific plant growth" (Steward 1933). Importantly, cultivation of important food plants (*Cyperus esculentus*), fish, and productive meadows were also supported through water redistribution by the *Nüümü* through extensive canals, berms, weirs, and dams to control water flow seasonally and yearly (Cheyenne Stone, Big Pine Paiute Tribe of the Owens Valley, personal communication, November 2024; Haverstock, Jayko, and Williams 2022; Lawton et al. 1976; Madley 2016). These agricultural canals were established between 639 and 512 years before the present day (Haverstock, Jayko, and Williams 2022), and they were documented across a distance of 57 miles along the Owens Valley, from Independence Creek (near the center of the OVPA) to Rock Creek (well north of the OVPA; Lawton et al. 1976). However, it is likely that agriculture through irrigation was not practiced near Owens Lake because food supply there depended on abundant fish and fly larvae supported by the lake (Lawton et al. 1976).

The Anglo incursion into Paiute territory was both unexpected and deadly (Varner 2009). In 1863, more than 900 Owens Valley Paiutes were forced to relocate to San Sebastian Reservation near Fort Tejon, and others fled into the mountains. When many returned in 1866, settlers had repurposed their fields for grazing and their irrigation ditches to grow introduced crops (Haverstock, Jayko, and Williams 2022; Lawton et al. 1976). By 1900, Inyo County's economy relied on agriculture, with some farms being owned by Native Americans, and many farms relying on the work of Native Americans (Borgias 2020). When the city of Los Angeles bought farmland in Owens Valley in the 1900s, many people, including the Paiute, lost ownership of, or access to, their water rights (Borgias 2024).[3] A top priority of some tribes is to restore the valley's hydrology by ending water diversions, export, and groundwater pumping because this water is the key resource that supports the diverse ecosystem types that have been degraded since water diversion (Cheyenne Stone, Big Pine Paiute Tribe of the Owens Valley, personal communication, November 2024; Sean Scruggs, Fort Independence Indian Reservation, personal communication, September 2024).

Of the approximately 19,000 residents of Inyo County, 13.8 percent identify as Native American, compared with 1.7 percent statewide (Cowan 2024). The Owens River is commonly used by Tribal members for hunting and fishing and gathering of native vegetation for various purposes, and the river area is used as a respite from extreme heat (Mel Joseph, Lone Pine Paiute-Shoshone Reservation, personal communication, August 2024). Tribal concerns have also been expressed for the conservation and protection of culturally important sites. The Patsiata Historical District was recently nominated for listing in the California Register of Historical Resources and in the National Register of Historic Places. The publicly available executive summary of the nomination notes that the District is associated with the traditions and history of the Indigenous People of *Payahuunadü* and that the archaeological sites provide important information about *Nüümü* and *Newe* history and culture.

SUMMARY

The geology and hydrology of Owens Valley conspire to create areas of sediment with the potential to be moved by wind and water. Vegetation cover plays an important role in minimizing dust emissions, and this vegetation is highly dependent on water availability. Diversion of surface water and decreases in groundwater have been linked to loss of meadows and decreases in shrub cover in the northern section of the OVPA. However, the area near Owens Lake has shown stable groundwater levels that have not been impacted by the diversion of Owens River nor groundwater pumping up-valley. Nevertheless, climate changes like extreme droughts could result in increased dust due to loss of vegetation and changes in groundwater if the area is not carefully managed. Additionally, climate change may impact the OVPA's vulnerability to dust emissions with more extreme and frequent precipitation events that will deposit more fine-grained materials and result in greater PM_{10} emissions.

---

[3] This sentence was edited after release of the report to clarify information from Borgias, 2024.

3

Sources of PM_{10} Emissions in the Owens Valley Planning Area

As described in the previous chapters, both on-lake and off-lake sources in the Owens Valley Planning Area (OVPA) continue to cause exceedances in particulate matter with an aerodynamic diameter of 10 micrometers or less, known as PM_{10} (Chapter 1). Potentially emissive sources include dunes, sand sheets, alluvial washes and fans, and sandy to silty flood deposits, in addition to on-lake features, which were more the focus of the prior National Academies of Sciences, Engineering, and Medicine report (see Chapter 2; NASEM 2020). This chapter evaluates the impacts of current and future emissions of off-lake sources to PM_{10} exceedances, the distribution of these sources in the OVPA, and how they might be better characterized and monitored.

Currently, there are two different indicator pollutants for particulate matter: $PM_{2.5}$, particles with an aerodynamic diameter of 2.5 micrometers or less, and PM_{10}, which by definition includes $PM_{2.5}$. The national health-based standard for PM_{10} is 150 micrograms per cubic meter ($\mu g/m^3$) averaged over 24 hours or 1 day. A PM_{10} National Ambient Air Quality Standard (NAAQS) exceedance occurs when an air quality monitoring station records a 24-hr average PM_{10} level over 150 $\mu g/m^3$. An exceedance day in the OVPA represents a day when one or more NAAQS PM_{10} monitors in the OVPA have an exceedance. Attainment with the PM_{10} NAAQS is achieved by having no more than one expected exceedance[1] per year at each monitor, averaged over 3 years (40 C.F.R. § 50.6[a]). Thus, four or more exceedances (at any level) at a single monitor in the OVPA within 3 years leads to nonattainment of the PM_{10} NAAQS. This means that it is not the level by which a measurement surpasses the standard that is important for an exceedance, but simply that it exceeds the standard. As of the end of 2024, the OVPA was in nonattainment of the federal PM_{10} NAAQS. Furthermore, OVPA is in nonattainment for the California state 24-hr average PM_{10} standard of 50 $\mu g/m^3$. However, as described in Chapter 1, the focus of this report is on the 24-hr NAAQS standard.

Unlike PM_{10}, $PM_{2.5}$ has two different primary standards as defined by the NAAQS: a 24-hour average of 35 $\mu g/m^3$ (assessed as the 98th percentile of these daily values for each year and then averaged over 3 years) and an annual average of 9 $\mu g/m^3$ for $PM_{2.5}$ (averaged over 3 years). Currently, most of the particulate matter monitors in the OVPA measure PM_{10}, not $PM_{2.5}$. It is generally understood that as dust-rich PM_{10} is controlled, the mineral fraction of $PM_{2.5}$ levels will be reduced as well, since during dust events a fraction (<20 percent) of PM_{10} is $PM_{2.5}$.

---

[1] Expected exceedance totals account for periods of time with missing data. As defined in 40 C.F.R. § 50.6(a), "In the simplest case, the number of expected exceedances at a site is determined by recording the number of exceedances in each calendar year and then averaging them over the past 3 calendar years. When data for a year are incomplete, it is necessary to compute an estimated number of exceedances for that year by adjusting the observed number of exceedances."

TRENDS IN OWENS LAKE AIR QUALITY OVER TIME

Since 2000, PM_{10} concentrations have been measured at nine shoreline and community sites around the lake (Figure 3-1), and these monitors are used to assess attainment with the PM_{10} NAAQS. Additional PM_{10} monitors are operated by the Fort Independence Indian Community of Paiute Indians and the Lone Pine Paiute Shoshone Tribe (both within the OVPA) but are not used to assess the OVPA's compliance with the PM_{10} NAAQS.

FIGURE 3-1 Current NAAQS PM_{10} monitoring sites in the OVPA. Hatch-markings indicate an area 3 km from the 3,600-ft-elevation regulatory shoreline.

TABLE 3-1 Exceedances of the PM_{10} 24-hr NAAQS at Monitors around Owens Lake from 2000 to 2023

Year	Area covered by DCM (square miles)	Annual PM_{10} Exceedances	Exceedance Day Count	Average 24-hr PM_{10} Exceedance ($\mu g/m^3$)	Maximum 24-hr PM_{10} Exceedance ($\mu g/m^3$)
2000	0.0	54	37	1,087	10,840
2001	9.4	81	46	1,413	20,750
2002	13.5	94	49	800	7,915
2003	14.8	72	37	1,115	16,619
2004	14.8	68	35	808	5,225
2005	18.9	49	28	627	3,988
2006	30.0	55	33	940	8,299
2007	30.0	18	14	272	727
2008	30.0	36	15	319	814
2009	30.0	49	19	339	1,506
2010	40.2	58	29	603	4,570
2011	40.2	52	24	641	13,380
2012	42.2	51	23	495	3,916
2013	42.2	20	13	283	529
2014	42.2	16	10	360	1,015
2015	45.3	30	14	337	1,487
2016	45.3	30	16	249	530
2017	49.0	46	17	411	2,164
2018	49.0	18	8	241	728
2019	49.0	10	5	234	387
2020	49.0	20	7	354	633
2021	49.0	15	11	257	605
2022	49.0	44	22	291	970
2023	49.2	28	16	337	861

NOTES: The annual PM_{10} exceedance total represents the number of times per year that an OVPA air quality monitoring station recorded a 24-hr average PM_{10} level over 150 $\mu g/m^3$. Exceedance day count is the number of distinct days during which *one or more* NAAQS PM_{10} monitor in the Owens Lake area experienced an exceedance of the 24-hour PM_{10} NAAQS standard of 150 $\mu g/m^3$. The data excludes wildfire events requested for exclusion in 2020 and 2021. Collocated daily PM_{10} exceedances at the Keeler location are counted once, represented by the maximum exceedance value. Only the original construction phase footprint was used for the area calculation and transition areas; reconstruction modification is not included. The total acreage includes 1.21 square miles of deferred areas, which contain eligible cultural resources and environmentally sensitive resources. DCM = dust control measures.
SOURCE: Area data from Arrash Agahi, Los Angeles Department of Water and Power (LADWP), personal communication, April 2025; Exceedance data from Ann Logan, Chris Howard, and Nik Barbieri, Great Basin Unified Air Pollution Control District (GBUAPCD), personal communication, July 2024 and January 2025.

Table 3-1 provides a summary of the PM_{10} exceedances of the NAAQS in the OVPA, and the extent of the historic lakebed area covered by dust control measures (DCMs) since 2000. The frequency and intensity of exceedances have decreased substantially as DCMs have been implemented on large areas of the lakebed.[2] During 2001–2003, which represents the first 3 years since the start of DCM implementation, there was an average of 44 (±6) exceedance days per year with an average 24-hr PM_{10} exceedance value of 1109 (±307) $\mu g/m^3$. By

---
[2] There is a statistically significant decreasing trend in the total number of exceedance days (p-value <0.01) and the 24-hr average PM_{10} exceedance value (p-value <0.01) shown in Table 3-1 using the Mann-Kendall test.

2021–2023, the average number of PM_{10} exceedance days per year decreased to 16 (±5) with almost a factor of four decrease in the average 24-hr PM_{10} exceedance value of 295 (±40) $\mu g/m^3$. Although in recent years the total number of days when the PM_{10} NAAQS was exceeded and the intensity of the exceedances have decreased sharply, the OVPA remains in nonattainment. Only two of the nine monitors—Lone Pine and Bill Stanley (hereafter, Stanley)—have consistently met the standard of no more than one exceedance per year, averaged over 3 years (Table 3-2).

By combining data on wind direction, wind speed, and the PM_{10} concentrations measured at the different monitoring sites on the lake, the Great Basin Unified Air Pollution Control District (the District) determined that by 2016 the contribution of on-lake sources to PM_{10} had decreased, and off-lake sources represented substantial relative contributions to PM_{10} exceedances (Figure 1-3; GBUAPCD 2016). However, in the 2016 State Implementation Plan (SIP) the District noted that emissions from off-lake sources were expected to decrease in the future as aeolian processes steadily removed dust from areas above the regulatory shoreline, which had historically originated on the lakebed (GBUAPCD 2016). The importance of this process, referred to as "winnowing," to future off-lake PM_{10} emissions is discussed further in Chapter 4. The following sections discuss the data and tools for understanding the trends in PM_{10} emissions from off-lake sources.

CURRENT TOOLS AND METHODS FOR PM_{10} SOURCE ATTRIBUTION

At Owens Lake, the District has used meteorological and particulate matter data as well as CALPUFF modeling for over two decades to identify the contributions of different dust source regions to observed PM_{10} concentrations (Ann Logan, GBUAPCD, personal communication, July 2023). Since 2017, a more systematic approach has been recorded in an exceedance database to track and categorize the sources for each exceedance, with there being five different established categories: 1) dust—primarily on-lake sources, 2) dust—primarily local off-lake sources, 3) dust—primarily regional event, 4) wildfire smoke, and 5) mixed—dust and wildfire sources. The classification of each event into the above categories is a nonquantitative best estimate and is based on the following data and components.

- **PM_{10} and meteorological data analysis.** PM_{10} pollution roses and time series of PM_{10} and wind speed for a few hours before and after an exceedance (Figure 3-2) are generated and analyzed. Hourly average wind speed and direction are computed from 5-second readings.

TABLE 3-2 Exceedances Per Year at Each PM_{10} Monitor in the OVPA Excluding Measurements with Exceptional Event Flags due to Wildfire Events

Year	2017	2018	2019	2020	2021	2022	2023	2024	Total
Dirty Socks	9	2	2	3	8	12	6	8	50
Keeler	7	2	2	4	2	9	7	5	38
Lizard Tail	9	3	1	3	1	2	2	1	22
Lone Pine	1	1	0	1	1	1	1	0	6
Mill Site	2	1	1	3	0	4	2	3	16
North Beach	5	3	2	1	2	2	2	0	17
Olancha	4	2	0	1	0	6	4	1	18
Shell Cut	8	4	2	3	1	8	4	6	36
Stanley	1	0	0	1	0	0	0	1	3
Total	46	18	10	20	15	44	28	25	206

NOTES: Table does not include two PM_{10} exceedances in 2020 and 14 exceedances in 2021, which were excluded due to wildfire smoke events. The Keeler site has three PM_{10} monitors; the highest annual exceedance count among the three monitors is listed.
SOURCE: Ann Logan, GBUAPCD, personal communication, May 2024; C. Howard, GBUAPCD, personal communication, April 2025.

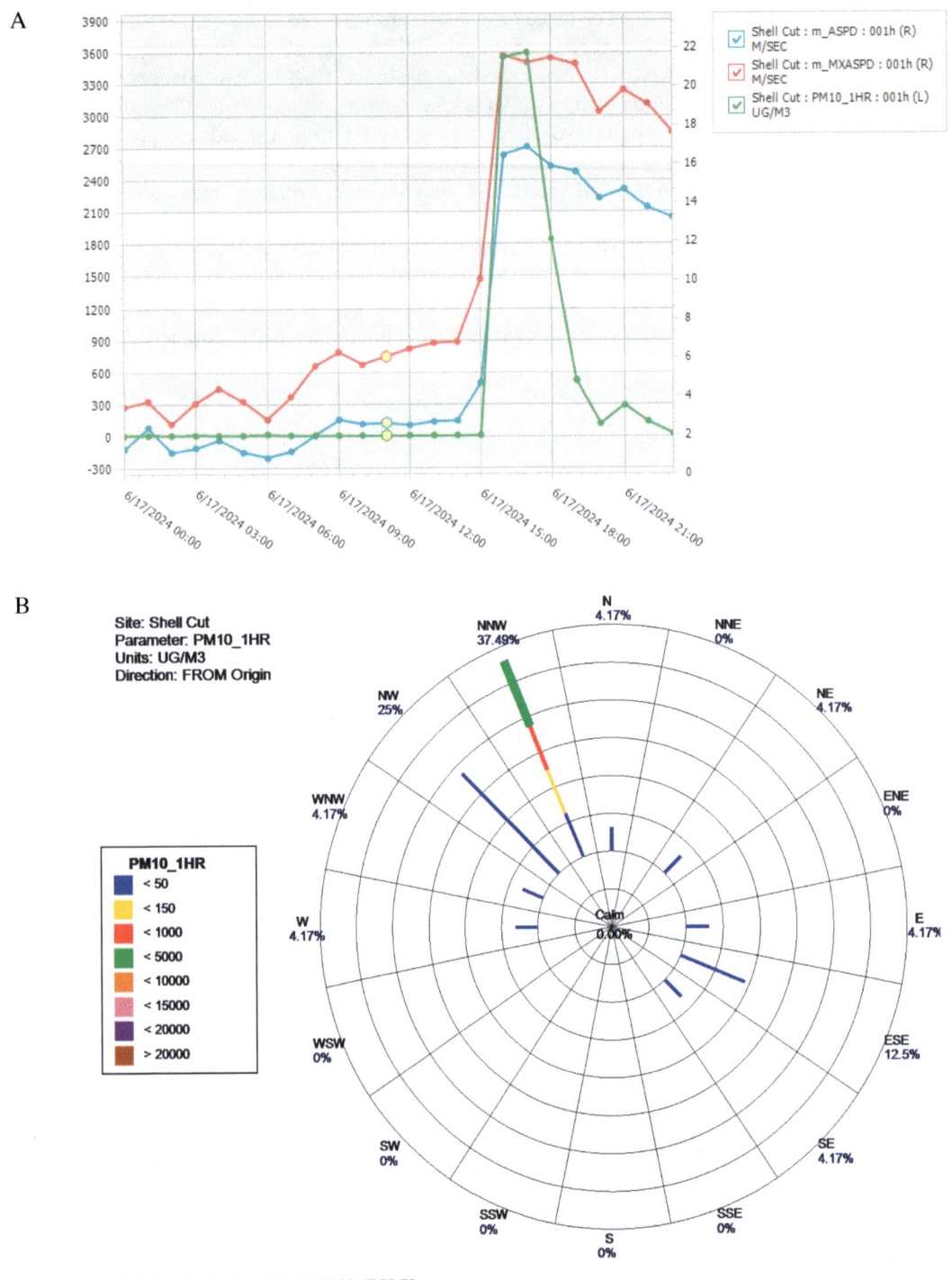

FIGURE 3-2 Example figures showing A) hourly PM_{10} (green, left axis, microgram per cubic meter [$\mu g/m^3$]), hourly wind speed (blue, right axis, m/s), and maximum wind speed over time (red, right axis, m/s) and B) a pollution rose of hourly PM_{10} measurements ($\mu g/m^3$) at Shell Cut during a single exceedance event, showing the direction of the PM_{10} source origin.
SOURCE: Chris Howard, GBUAPCD, personal communication, July 2024.

- **CALPUFF modeling.** Hourly dispersion of on-lake PM_{10} plumes is modeled using CALPUFF, driven by CALMET-derived meteorological (e.g., wind) fields, and estimates of sand flux. CALPUFF is a Lagrangian, non-steady state, source-oriented model that follows air parcels emitted from the modeled source regions as they undergo horizontal and vertical diffusion during transport (Scire, Strimaitis, and Yamartino 2000). The District uses CALMET (Scire et al. 2000) to provide meteorological inputs to CALPUFF on an hourly basis. CALMET is a diagnostic meteorological model that develops meteorological fields (e.g., wind, temperature, mixing height, Monin Obukhov length, friction velocity, and others) using meteorological measurements, orography, and land use data. Sand flux estimates generated from sand motion monitoring (Gillette, Ono, and Richmond 2004; Ono 2006) on the lakebed and Keeler Dunes are used together with finer-scale wind data across the lake and PM_{10} measurements at upwind and downwind sites to identify the predominant area(s) responsible for the exceedance.
- **Review of photographic evidence.** If the exceedance occurs during the daytime, time-lapse photos from 20 cameras sited in 11 locations distributed around the lake and in the nearby foothills are reviewed to assess whether the source of dust is regional or more localized.
- **$PM_{2.5}$.** Using $PM_{2.5}$ data from the Keeler monitoring station, ratios of $PM_{2.5}$ to PM_{10} can be used to determine the influence of wildfire smoke on elevated PM_{10}. Research has shown that wildfire smoke has a larger fraction of aerosols in the $PM_{2.5}$ fraction (Schweizer, Cisneros, and Buhler 2019), and therefore exceedance events with a very high $PM_{2.5}$ fraction (>70 percent) are determined to be driven by a wildfire instead of wind-blown dust. Because wildfire events deteriorate air quality on a regional scale, this source attribution at the Keeler monitoring station is assumed to apply to all other monitoring locations in the OVPA.
- **Observation and news.** Finally, notes from site operators and field inspectors along with weather reports on larger-scale synoptic events or media reports of dust storms are considered to qualify the exceedance event as regional, on-lake, or local off-lake and are also used to identify local anthropogenic sources (e.g., construction; Ann Logan, GBUAPCD, personal communication, July 2024).

Because exceedances are rare events, they may not be robust indicators of the emission potentials of sources. However, this approach of exceedance source attribution is systematic and detailed enough to parse out events as having on- or off-lake sources and differentiate the primary source of local off-lake sources (e.g., flood deposits, landfills, regional dust events, wildfires, dunes; see Box 3-1). Some ambiguity remains when several sources line up with the dominant wind direction or if the wind direction is near the angle used to separate on- and off-lake

---

BOX 3-1
Classification of Sources in the Exceedance Database

Based on the methods described above for source attribution (e.g., photographic evidence, modeling), the District describes potential sources for each exceedance event in the exceedance database. As shown throughout this chapter, the panel used the District's descriptions in the exceedance database to identify a list of likely sources for each exceedance event and then summarized the most common sources for each monitor and for off-lake sources generally. The District describes many sources as "flash-flood deposits" in the exceedance database, including those that are 1) channelized flood deposits; 2) sheet/overland flow deposits; and 3) deposits from impounded floodwaters (Chris Howard, GBUAPCD, personal communication, December 2024). The panel did not attempt to re-classify every "flash-flood deposit" as a specific type of flood deposit in the exceedance database, but it does attempt to be specific in the text where evidence is available. Where evidence is not available to be more specific, the panel refers to these events generally as "flood deposits." In contrast, the District uses the term "alluvial fans" to refer to a general landscape feature and not a specific flood event.

source attribution. Additionally, categorization is somewhat uncertain when multiple sources of dust are suspected to contribute to an event—for example, when exceedance events driven primarily by on-lake sources are also impacted by some amount of PM_{10} emission from off-lake sources. For example, in their modeling of PM_{10} events driven by on-lake sources, the District includes a fixed background concentration in their base model runs or a variable background in their hybrid approach to account for contributions from up-valley or other off-lake sources.

IMPACTS OF LOCAL OFF-LAKE SOURCES TO PM_{10} EXCEEDANCES

In this section, the panel examines data from individual air quality monitoring stations in the OVPA to identify the most frequent local off-lake sources of PM_{10}. In recent years, five monitoring stations show a large number of PM_{10} exceedances from local off-lake sources—Dirty Socks, Keeler, Mill, Olancha, and Shell Cut (Figures 3-3 and 3-4). Stanley, Lizard Tail, and Lone Pine sites have had fewer PM_{10} exceedances due to local off-lake sources. For this reason, Stanley and Lizard Tail will not be analyzed further in this chapter. However, Lone Pine is analyzed in addition to data from Fort Independence to examine the potential influence of local off-lake sources from the valley north of Owens Lake. The following sections discuss the data available for the remaining seven monitoring sites (in order from south to north) with an emphasis on exceedances and potential local off-lake sources for each of these sites.

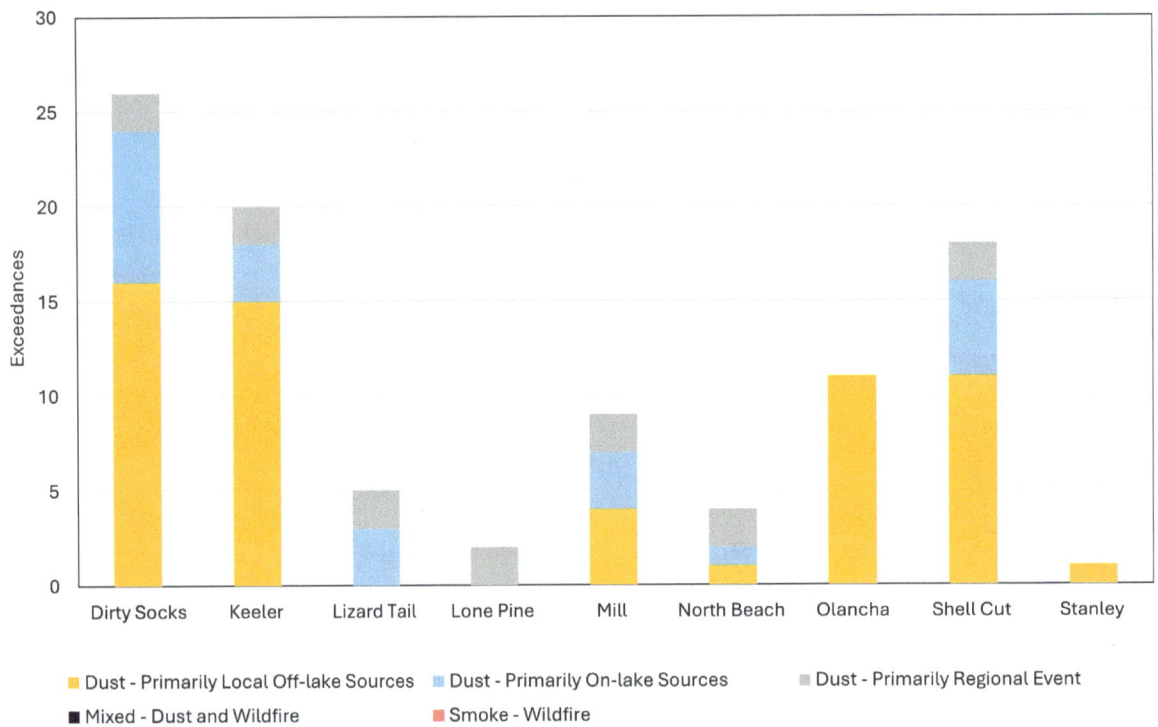

FIGURE 3-3 PM_{10} exceedances by site from 2022 to 2024, as categorized in the District's exceedance database.
NOTES: These data represent the last 3 years, the period over which NAAQS design value[a] compliance is calculated. Compliance with the PM_{10} NAAQS is achieved by having no more than one exceedance per year at each monitor, averaged over 3 years. No "Smoke—Wildfire" or "Mixed—Dust and Wildfire" events were recorded over this 3-year timeframe.
SOURCE: Data from exceedance database, Chris Howard, GBUAPCD, personal communication, August 2024 and April 2025.

---

[a] Design values are used to determine the needed level of control to demonstrate attainment of the PM_{10} NAAQS. The design value is determined by calculating the expected exceedances each year, for 3 calendar years according to the approach discussed in 40 C.F.R. § 50, appendix K. This calculation adjusts for the number of days in the year without PM_{10} data.

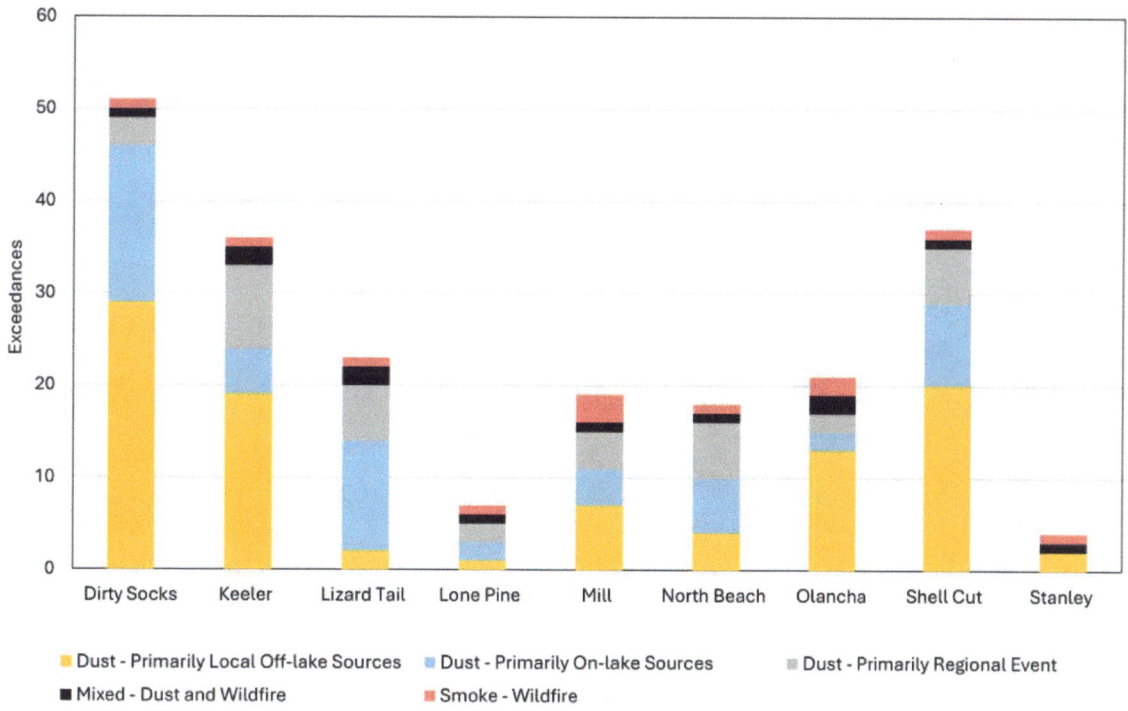

FIGURE 3-4 PM_{10} exceedances by site from 2017 to 2024.
NOTES: These dates represent the full dataset since 2017, when a systematic approach began to categorize the sources for each exceedance.
SOURCE: Data from exceedance database, Chris Howard, GBUAPCD, personal communication, August 2024 and April 2025.

Olancha Monitoring Site

The Olancha monitoring site is currently located near Olancha Creek, approximately 0.3 miles (500 m) west of Highway 395. It was previously located approximately 2.2 km to the southeast and moved to its present location in 2019 (Figure 3-5A). The current location of the Olancha monitoring station to the west of the Olancha Dunes complex is not ideal for characterizing emissions sourced from the dunes, given the bimodal north–south orientation of the wind regime.

Based on an analysis using consistent calculations over time, on-lake sources dominated exceedances at the Olancha monitoring site until 2020, but since 2020, PM_{10} exceedances have been mostly associated with off-lake sources (Figure 3-6). Analyses of the data in Figure 3-6 show a statistically significant decreasing trend (Mann-Kendall test p-value <0.01) in the number of consistently calculated PM_{10} exceedances from on-lake sources, while there is no significant trend in the number of consistently calculated off-lake exceedances at Olancha (Mann-Kendall test p-value >0.05). Extremely high hourly concentrations of PM_{10} have been observed from the west where there have been sheet/overland flow deposits, instead of from the lake to the north or from the Olancha Dunes to the east (Figure 3-5B). The PM_{10} hourly concentrations during some of the dust events in 2021–2023 have reached 10,000 $\mu g/m^3$.

Based on the panel's tally of sources described in the District's exceedance database, it is apparent that approximately 62 percent of the local off-lake PM_{10} exceedances between 2017 and 2024 were fully or partially attributed to flood deposits located west, southwest, or south of the site, approximately 38 percent to disturbed surfaces due to human activity (e.g., road construction), and only around 15 percent to Olancha Dunes (Figure 3-7). In the exceedance database, the District has not identified specific sources within the Olancha Dunes, such as the active off-highway vehicle (OHV) riding and dispersed camping recreational area that makes up approximately 36 percent of the Olancha Dunes area. It is well documented at other managed OHV recreation sites in southern

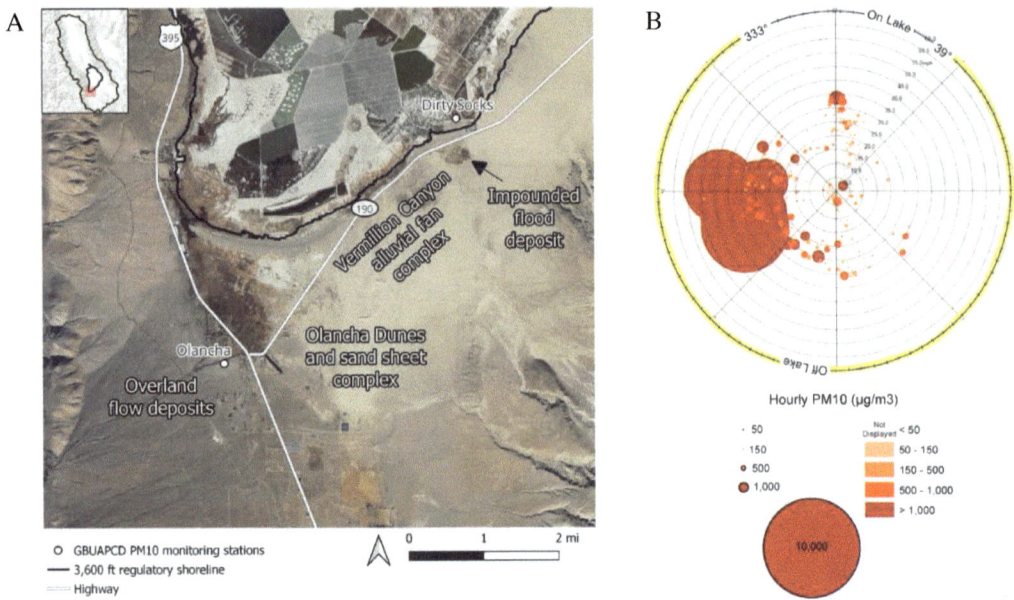

FIGURE 3-5 A) Location of current Olancha monitoring site relative to potential local off-lake sources. B) A pollution rose plot based on 2021–2023 hourly PM_{10} data at the Olancha monitoring site.
NOTES: B) The markers, sized and colored by the hourly PM_{10} concentrations for data points above 50 $\mu g/m^3$, are mapped out on a polar plot. The angular position on this plot represents the direction from which the wind was blowing at hourly intervals at the site, while the radial distance from the center of the polar plot represents the hourly wind speed in miles per hour. Wildfire smoke events have been removed.
SOURCE: A) Adapted from Google Earth; B) Generated by GBUAPCD.

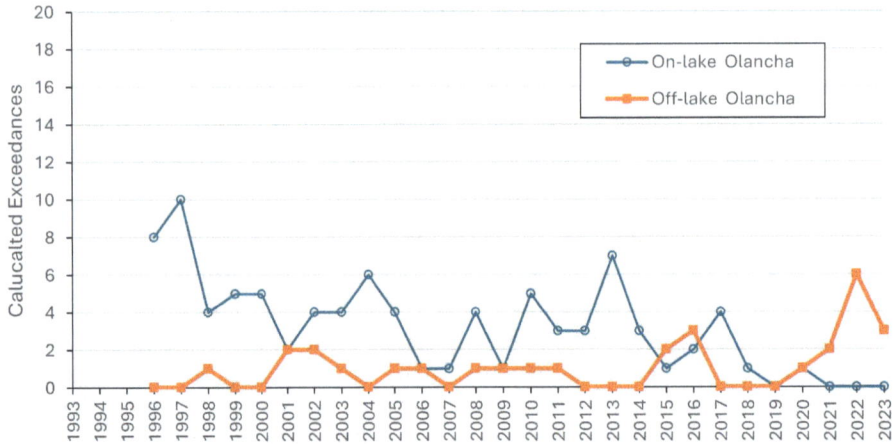

FIGURE 3-6 On-lake and off-lake exceedances at the Olancha monitoring site, with calculations based on consistent metrics over time.
NOTES: These data do not reflect official exceedance counts, but this analysis was conducted to understand long-term trends using consistent metrics. Exceedance counts and on/off-lake attribution based on screened wind directions and a threshold of 150 $\mu g/m^3$ for exceedances. For all years, on/off-lake exceedance counts were purely classified by screened wind directions and may include wildfire smoke events and regional events. For all hours without data when the wind was coming from an opposite direction (on- or off-lake), a background concentration of 20 $\mu g/m^3$ was assumed. The location of the monitor was moved in 2019, as described in the text.
SOURCE: Data from Chris Howard, GBUAPCD, personal communication, November 2024.

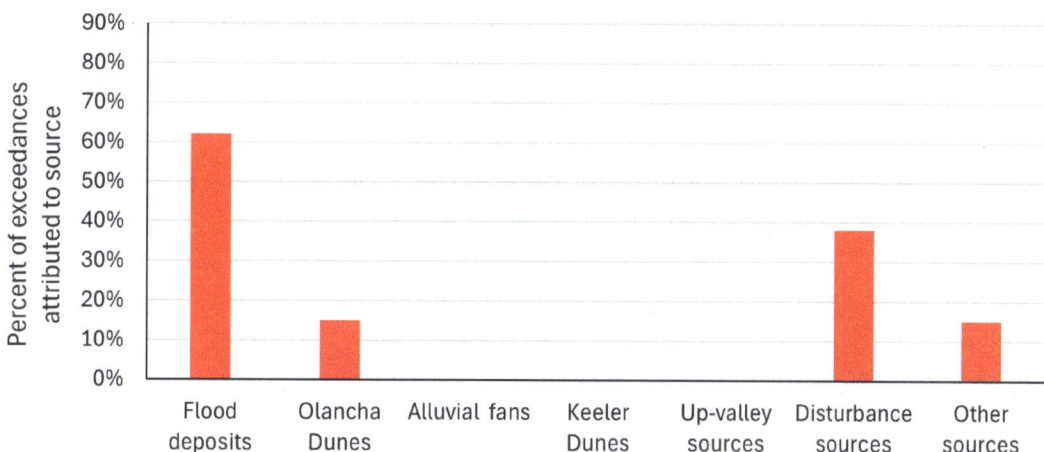

FIGURE 3-7 Summary of the sources of local off-lake PM_{10} exceedances at Olancha monitoring site during 2017–2024.
NOTES: A single exceedance event may be partially attributed to multiple sources. "Other sources" include one event attributed to "uncontrolled surfaces" and one attributed to a source west of Olancha. Emissions from the Olancha Dunes are better measured at the Dirty Socks monitoring site.
SOURCE: Based on data in the GBUAPCD exceedance database, Chris Howard, GBUAPCD, personal communication, August 2024 and April 2025.

California (e.g., Imperial Dunes, Oceano Dunes) that OHV activities contribute to declines in vegetation cover, disturbance of surface conditions, and elevated exceedances of PM_{10} standards. Although research on these associations is lacking at the Olancha Dunes, it is reasonable to assume that areas in the Owens Valley subject to OHV activities could also contribute to elevated PM_{10} emissions in the region. Increased monitoring and baseline research would help identify the nature and extent of land use change and potential contributions of OHV activity to dust emissions at Olancha Dunes.

Dirty Socks Monitoring Site

The Dirty Socks monitor is located along the southeastern edge of Owens Lake (Figure 3-8A). During 2017–2024, of all the monitors in the OVPA, Dirty Socks had the largest number of exceedances attributed to local off-lake sources (Figure 3-4). As shown in an analysis using consistent exceedance calculations over time, on-lake sources dominated at the Dirty Socks monitoring site until 2012, but since 2015 (when monitoring at the site resumed), the frequency of consistently calculated PM_{10} exceedances due to off-lake sources have been greater than or equal to on-lake sources (Figure 3-9). Overall, there is a statistically significant decreasing trend (Mann-Kendall test p-value <0.01) in the number of consistently calculated PM_{10} exceedances from on-lake sources while there is no significant trend in the number of calculated off-lake exceedances at Dirty Socks (Mann-Kendall test p-value >0.05). Analysis of wind direction and hourly PM_{10} data from 2021–2023 (Figure 3-8B) reveal off-lake sources are mostly to the south of the monitor toward the Vermillion Canyon Alluvial Fan Complex (with flood deposits) and Olancha Dunes. Furthermore, hourly PM_{10} concentrations during the largest off-lake dust events are slightly lower than the largest on-lake events, but events with hourly PM_{10} greater than 1,000 $\mu g/m^3$ occurred more frequently from off-lake sources than on-lake.

Among the 29 PM_{10} exceedances from local off-lake sources during 2017–2024 reported in the District's exceedance database, 59 percent were at least partially attributed to flood deposits, 48 percent to Olancha Dunes, 10 percent to alluvial fans, and 17 percent to other sources (Figure 3-10).[3] As discussed in the previous section,

---

[3] Total exceeds 100 percent because a single exceedance may be attributed to more than one source.

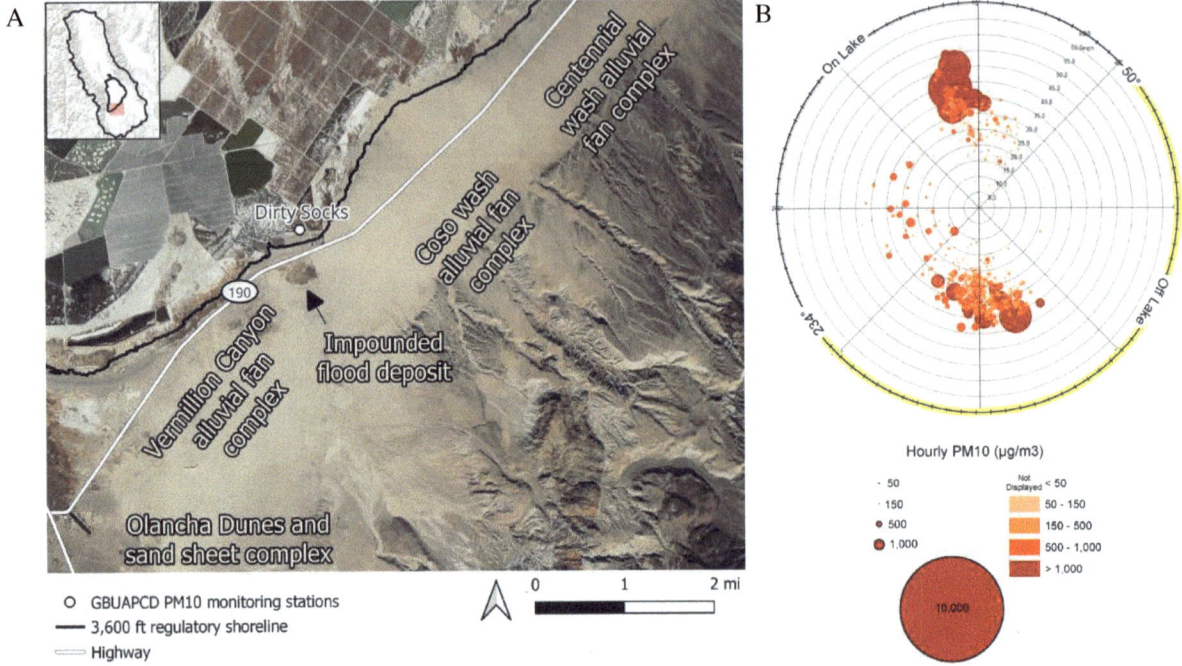

FIGURE 3-8 A) Location of Dirty Socks monitoring site relative to potential local off-lake sources. B) Hourly PM_{10} pollution rose plots at Dirty Socks for 2021–2023.
NOTES: B) The markers, sized and colored by the hourly PM_{10} concentrations for data points above 50 $\mu g/m^3$, are mapped out on a polar plot. The angular position on this plot represents the direction from which the wind was blowing at hourly intervals at the site, while the radial distance from the center of the polar plot represents the hourly wind speed. Wildfire smoke events have been removed.
SOURCES: A) Adapted from Google Earth; B) GBUAPCD.

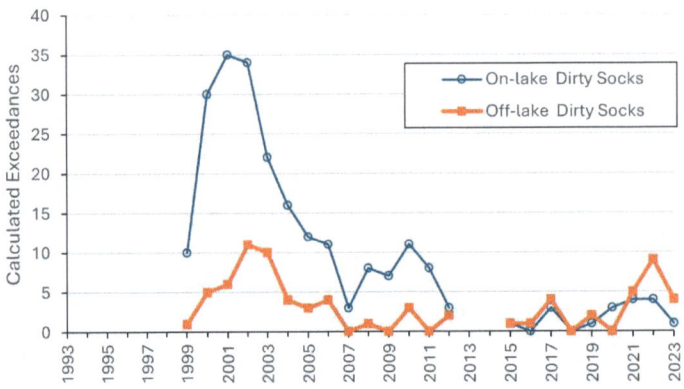

FIGURE 3-9 On-lake and off-lake exceedances at the Dirty Socks monitoring site, with calculations based on consistent metrics over time.
NOTES: These data do not reflect official exceedance counts, but this analysis was conducted to understand long-term trends using consistent metrics. Exceedance counts and on/off-lake attribution based on screened wind directions and a threshold of 150 $\mu g/m^3$ for exceedances. For all years, on/off-lake exceedance counts were purely classified by screened wind directions and may include wildfire smoke events and regional events. For all hours without data when the wind was coming from an opposite direction (on- or off-lake), a background concentration of 20 $\mu g/m^3$ was assumed. Data were not collected between December 2012 and January 2015.
SOURCE: Data from Chris Howard, GBUAPCD, personal communication, November 2024.

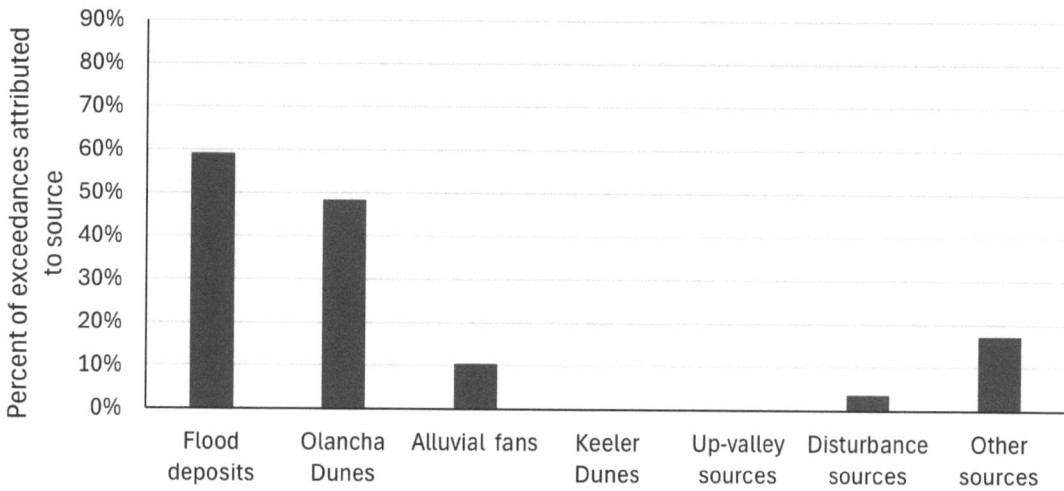

FIGURE 3-10 Summary of sources of local off-lake PM_{10} exceedances at Dirty Socks monitoring site during 2017–2024.
NOTES: A single exceedance event may be partially attributed to multiple sources. "Other sources" include two events attributed to unknown sources, two attributed to shoreline deposits, and one attributed to open desert.
SOURCE: Exceedance database, Chris Howard, GBUAPCD, personal communication, August 2024 and April 2025.

there is not sufficient evidence to suggest how much material is being sourced from the OHV and dispersed camping recreational area that makes up approximately 36 percent of the Olancha Dunes area.

Shell Cut Monitoring Site

The Shell Cut monitoring site is located roughly at the midpoint along the southeastern edge of Owens Lake (Figure 3-11A). An analysis using consistent exceedance metrics over time shows that from 2001 to 2009, consistently calculated PM_{10} exceedances were predominantly due to dust events from on-lake sources, but a majority of calculated PM_{10} exceedances since 2011 have been attributed to off-lake sources (Figure 3-12). Overall there is a statistically significant decreasing trend (Mann-Kendall test p-value <0.01) in the number of consistently calculated PM_{10} exceedances from on-lake sources while there is no significant trend in the number of calculated off-lake exceedances at Shell Cut (Mann-Kendall test p-value >0.05). Analysis of wind direction and hourly PM_{10} data from 2021 to 2023 (Figure 3-11B) reveal off-lake sources to the south towards Coso Wash Alluvial Fan Complex. Extensive flood deposits from the remnants of Hurricane Kay in 2022 have been mapped near the Centennial Wash, but these do not appear as major sources on the rose diagram (Figure 3-11B). PM_{10} hourly concentrations during 2021–2023 events were generally comparable for on-lake and off-lake sources, except for one on-lake dust event that resulted in the hourly concentration of approximately 10,000 $\mu g/m^3$ PM_{10} (Figure 3-11B).

Of a total of 21 local off-lake PM_{10} exceedances that were reported in 2017–2024 in the District's exceedance database, 90 percent of these had at least partial contributions from flood deposits, 20 percent from alluvial fans, and 20 percent from other sources (e.g., shoreline deposits; Figure 3-13).

Mill Monitoring Site

The Mill monitoring site is located along the eastern edge of the Owens Lake bed (Figure 3-14A). The number of exceedances at both on- and off-lake sites are generally low (Figures 3-2 and 3-3), and an analysis of consistent exceedance metrics shows no clear trend over time (Mann-Kendall test p-value >0.05; Figure 3-15). Off-lake dust events affecting the Mill site are most often associated with winds from the southeast, with hourly PM_{10}

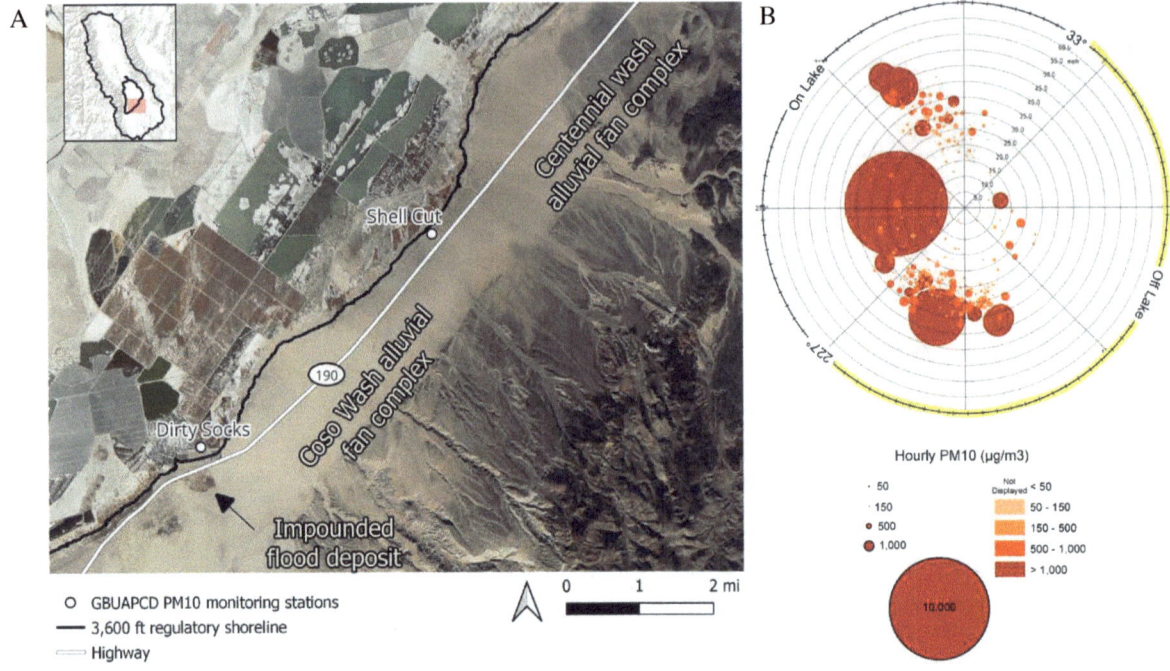

FIGURE 3-11 A) Location of Shell Cut monitoring site relative to potential local off-lake sources. B) A pollution rose plot based on 2021–2023 hourly PM_{10} data at the Shell Cut monitoring site.
NOTES: A) Some highway berms and ditches exist in this area but are not mapped because they were not analyzed in detail by the panel. B) The markers, sized and colored by the hourly PM_{10} concentrations for data points above 50 $\mu g/m^3$, are mapped out on a polar plot. The angular position on this plot represents the direction from which the wind was blowing at hourly intervals at the site, while the radial distance from the center of the polar plot represents the hourly wind speed. Wildfire smoke events have been removed.
SOURCES: A) Adapted from Google Earth; B) GBUAPCD.

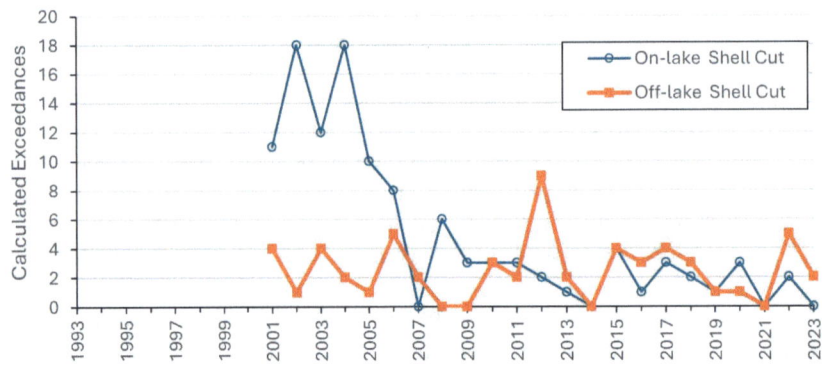

FIGURE 3-12 On-lake and off-lake exceedances at the Shell Cut monitoring site, with calculations based on consistent metrics over time.
NOTES: These data do not reflect official exceedance counts, but this analysis was conducted to understand long-term trends using consistent metrics. Exceedance counts and on/off-lake attribution based on screened wind directions and a threshold of 150 $\mu g/m^3$ for exceedances. For all years, on/off-lake exceedance counts were purely classified by screened wind directions and may include wildfire smoke events and regional events. For all hours without data when the wind was coming from an opposite direction (on- or off-lake), a background concentration of 20 $\mu g/m^3$ was assumed.
SOURCE: Data from Chris Howard, GBUAPCD, personal communication, November 2024.

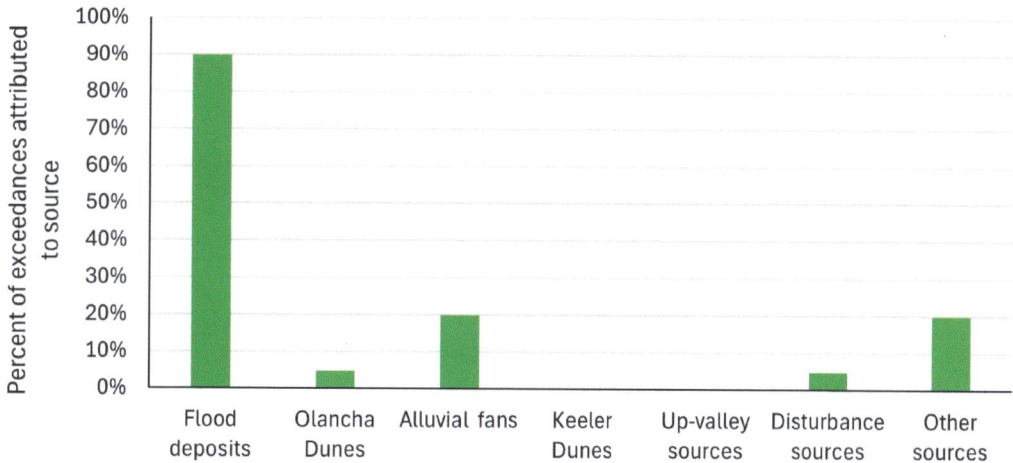

FIGURE 3-13 Summary of sources of local off-lake PM$_{10}$ exceedances at Shell Cut monitoring site during 2017–2024.
NOTES: Note that a single exceedance event may be partially attributed to multiple sources. "Other sources" include two events attributed to shoreline deposits, one attributed to regional sources, and one attributed to Coso.
SOURCE: Data from exceedance database, Chris Howard, GBUAPCD, personal communication, August 2024 and April 2025.

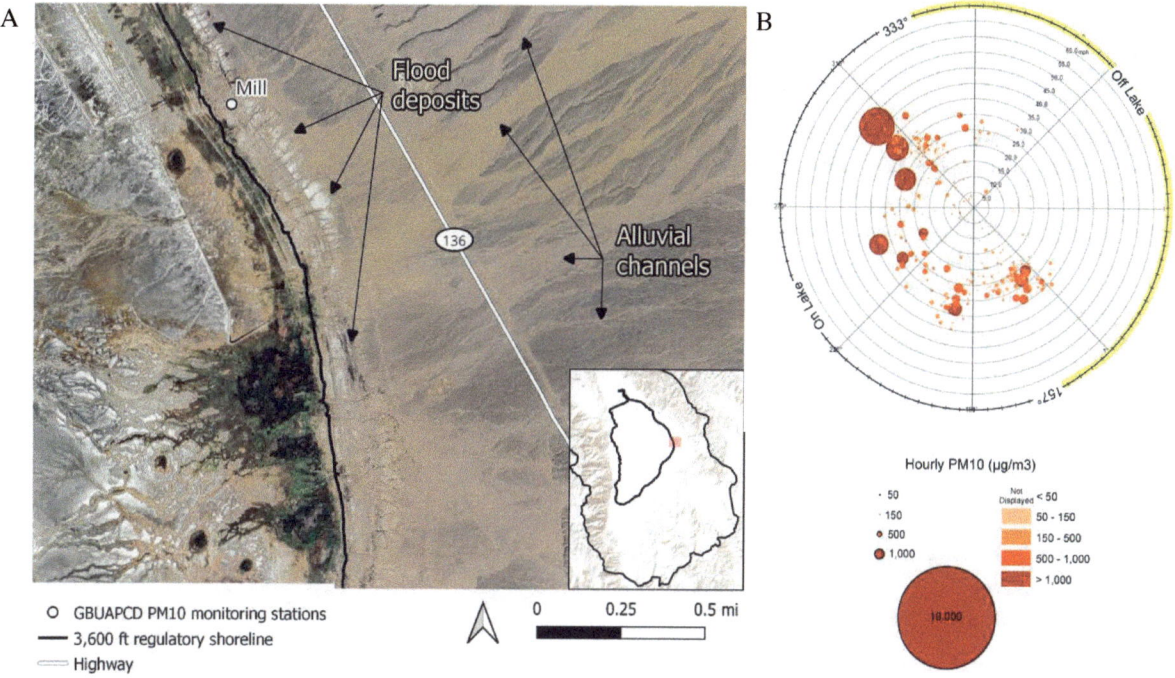

FIGURE 3-14 A) Location of the Mill monitoring site relative to potential local off-lake sources.
B) A pollution rose plot based on 2021–2023 hourly PM$_{10}$ data at the Mill monitoring site.
NOTES: A) Highway berms exist in this image but are not mapped as they were not analyzed by the panel at this location.
B) The markers, sized and colored by the hourly PM$_{10}$ concentrations for data points above 50 μg/m^3, are mapped out on a polar plot. The angular position on this plot represents the direction from which the wind was blowing at hourly intervals at the site, while the radial distance from the center of the polar plot represents the hourly wind speed. Wildfire smoke events have been removed.
SOURCES: A) Adapted from Google Earth. B) GBUAPCD.

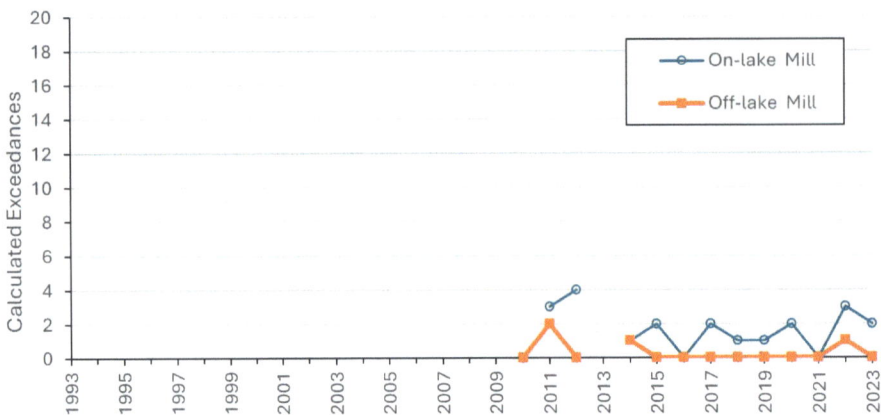

FIGURE 3-15 On-lake and off-lake exceedances at the Mill monitoring site, with calculations based on consistent metrics over time.
NOTES: These data do not reflect official exceedance counts, but this analysis was conducted to understand long-term trends using consistent metrics. Exceedance counts and on/off-lake attribution based on screened wind directions and a threshold of 150 $\mu g/m^3$ for exceedances. For all years, on/off-lake exceedance counts were purely classified by screened wind directions and may include wildfire smoke events and regional events. For all hours without data when the wind was coming from an opposite direction (on- or off-lake), a background concentration of 20 $\mu g/m^3$ was assumed. Data were not collected between December 2012 and August of 2014.
SOURCE: Data from Chris Howard, GBUAPCD, personal communication, November 2024.

concentrations between 500 and 1000 $\mu g/m^3$. This direction is consistent with the location of recent flood deposits, including those mapped from the remnants of Hurricane Kay in 2022. In fact, for the seven PM_{10} exceedances attributed to local off-lake sources in 2017–2024 in the District's exceedance database, 57 percent had at least partial contributions from flood deposits, 29 percent from Keeler Dunes, 14 percent from up-valley sources, and 29 percent from other sources (Figure 3-16).

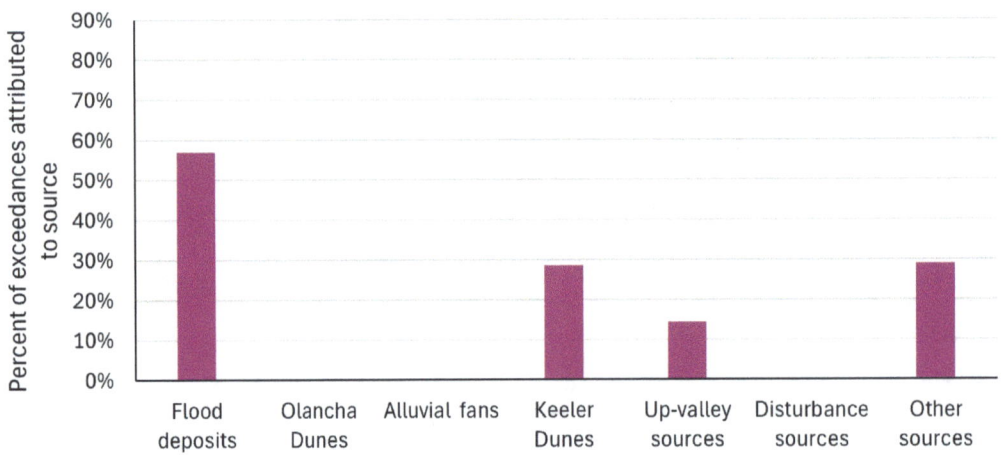

FIGURE 3-16 Summary of sources of local off-lake PM_{10} exceedances at Mill monitoring site during 2017–2024.
NOTES: Note that a single exceedance event may be partially attributed to multiple sources. "Other sources" includes one event attributed to a regional source and one attributed to "shoreline pockets."
SOURCE: Data from exceedance database, Chris Howard, GBUAPCD, personal communication, August 2024 and April 2025.

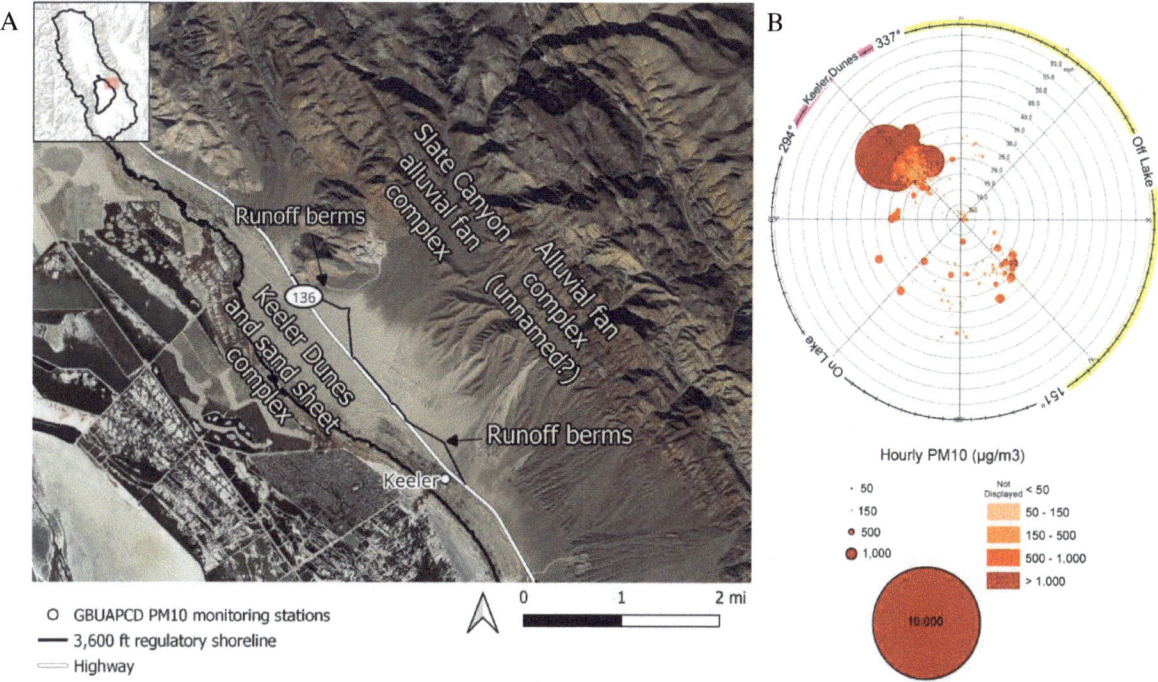

FIGURE 3-17 A) Location of Keeler monitoring site relative to potential local off-lake sources. B) A PM_{10} pollution rose plot based on 2021–2023 hourly data at the Keeler monitoring site.
NOTES: B) The markers, sized and colored by the hourly PM_{10} concentrations for data points above 50 $\mu g/m^3$, are mapped out on a polar plot. The angular position on this plot represents the direction from which the wind was blowing at hourly intervals at the site, while the radial distance from the center of the polar plot represents the hourly wind speed. Wildfire smoke events have been removed.
SOURCES: A) Adapted from Google Earth. B) GBUAPCD.

Keeler Monitoring Site

The Keeler monitoring site, located east-northeast of Owens Lake (Figure 3-17A), has long been a concern for its high levels of PM_{10} emissions from off-lake sources. As shown in an analysis of data from 1993–2023 based on consistent exceedance metrics (Figure 3-18), there is a statistically significant decreasing trend (Mann-Kendall test p-value <0.01) in the number of consistently calculated PM_{10} exceedances from on-lake sources while there is no significant trend in the number of calculated off-lake exceedances at Keeler over the full monitoring period (Mann-Kendall test p-value >0.05). Recently, a few years have had historically low numbers of consistently calculated exceedances after implementation of the Keeler Dunes Dust Control Project (Box 6-1) project, although the apparent downward trend in calculated exceedances since 2014 has not been sustained in 2022 and 2023 (Figure 3-18).

Many elevated hourly PM_{10} concentrations in 2021–2023 are associated with winds from the direction of Keeler Dunes or winds from the southeast in the direction of some flood deposits that have been diverted by runoff berms above Highway 136 (Figure 3-17B). The hourly PM_{10} values from 2021–2023 under the influence of winds from the direction of Keeler Dunes were significantly higher than values observed under the influence of wind from other on-lake or off-lake directions, approaching and occasionally exceeding thousands of $\mu g/m^3$ (Figure 3-17).[4]

---

[4] Based on the Student t-test, committee analysis showed that hourly PM_{10} values in 2021–2023 under the influence of winds from the direction of Keeler Dunes were significantly higher (average value of 64.6±244 $\mu g/m^3$) than the values observed under the influence of wind from other off-lake (not including the Keeler Dunes) or on-lake directions (average value of 17±33 $\mu g/m^3$).

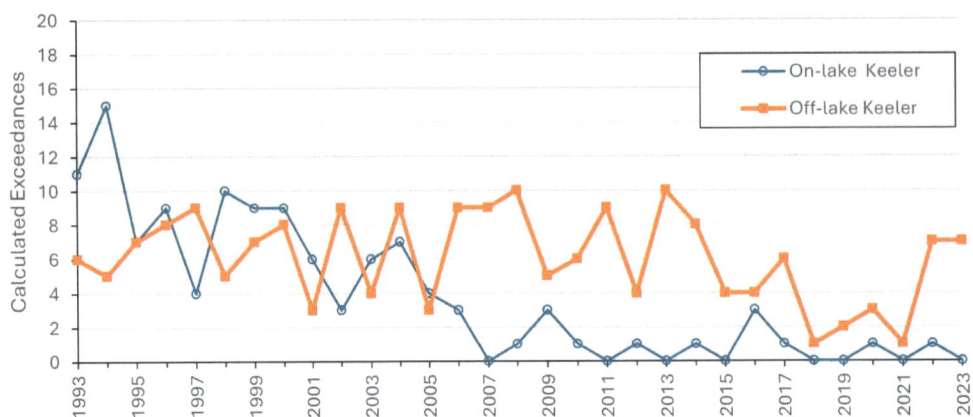

FIGURE 3-18 On-lake and off-lake exceedances at the Keeler monitoring site, with calculations based on consistent metrics over time.
NOTES: These data do not reflect official exceedance counts, but this analysis was conducted to understand long-term trends using consistent metrics. Exceedance counts and on/off-lake attribution based on screened wind directions and a threshold of 150 µg/m³ for exceedances. For all years, on/off-lake exceedance counts were purely classified by screened wind directions and may include wildfire smoke events and regional events. For all hours without data when the wind was coming from an opposite direction (on- or off-lake), a background concentration of 20 µg/m³ was assumed. The screened wind directions for off-lake exceedances include the Keeler Dunes.
SOURCE: Data from Chris Howard, GBUAPCD, personal communication, November 2024.

Among the 19 local off-lake PM_{10} exceedances in 2017–2024 reported in the District's exceedance database, 79 percent had at least partial contributions from Keeler Dunes, 21 percent from flood deposits, 16 percent from up-valley sources, and 21 percent from other sources (including 1 from Swansea Dunes; Figure 3-19). Since Keeler Dunes, Swansea Dunes, and flood deposits around the dunes are all in the same general direction as the Keeler monitoring site, distinguishing the influence of each of these sources to a PM_{10} exceedance event at Keeler is challenging and remains uncertain.

Continuous $PM_{2.5}$ measurements have been conducted at the Keeler monitoring site since 2009. These data can be useful to identify cases where the OVPA is under the influence of regional wildfires since the coarse fraction of PM_{10} (i.e., $[PM_{10}-PM_{2.5}]/PM_{10}$) is expected to be less than 30 percent under these conditions (Schweizer, Cisneros, and Buhler 2019). One may also expect different PM coarse fractions from different dust sources of PM_{10}. Upon committee examination of the fractional distribution of PM coarse fractions observed at the Keeler site from 2015–2023 under the influence of air masses from on-lake sources, off-lake (not including Keeler-Dune sources), and off-lake from Keeler Dunes, it is apparent that there is an increased contribution of coarse particles to PM_{10} with Keeler Dunes' emissions. However, such high coarse fractions are not unique to Keeler Dunes because similar values are occasionally also observed under the influence of other on-lake or off-lake sources (Figure 3-20). This observation makes it difficult to use measurements of $PM_{2.5}$ to unambiguously distinguish between on-lake and off-lake dust sources of PM_{10}.

North Beach Monitoring Site

At the North Beach monitor, located at the northern end of Owens Lake (Figure 3-21A), the majority of exceedances since 2017 have been attributed to sources other than the lakebed (Figure 3-22). Based on an analysis of data using consistent exceedance metrics over time, there is a statistically significant decreasing trend (Mann-Kendall test p-value <0.01) in the number of consistently calculated PM_{10} exceedances from on-lake sources while there is no significant trend in the number of calculated off-lake exceedances at North Beach (Mann-Kendall

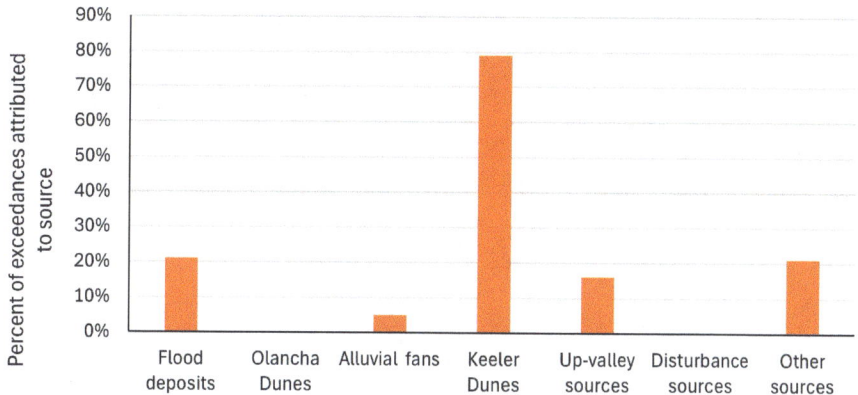

FIGURE 3-19 Summary of sources of local off-lake PM$_{10}$ exceedances at Keeler monitoring site during 2017–2024. NOTES: Note that a single exceedance event may be partially attributed to multiple sources. "Other sources" includes one event attributed to a source north of Lizard Tail, one attributed to the Swansea Dunes, one regional event, and one unknown.
SOURCE: Data from exceedance database, Chris Howard, GBUAPCD, personal communication, August 2024 and April 2025.

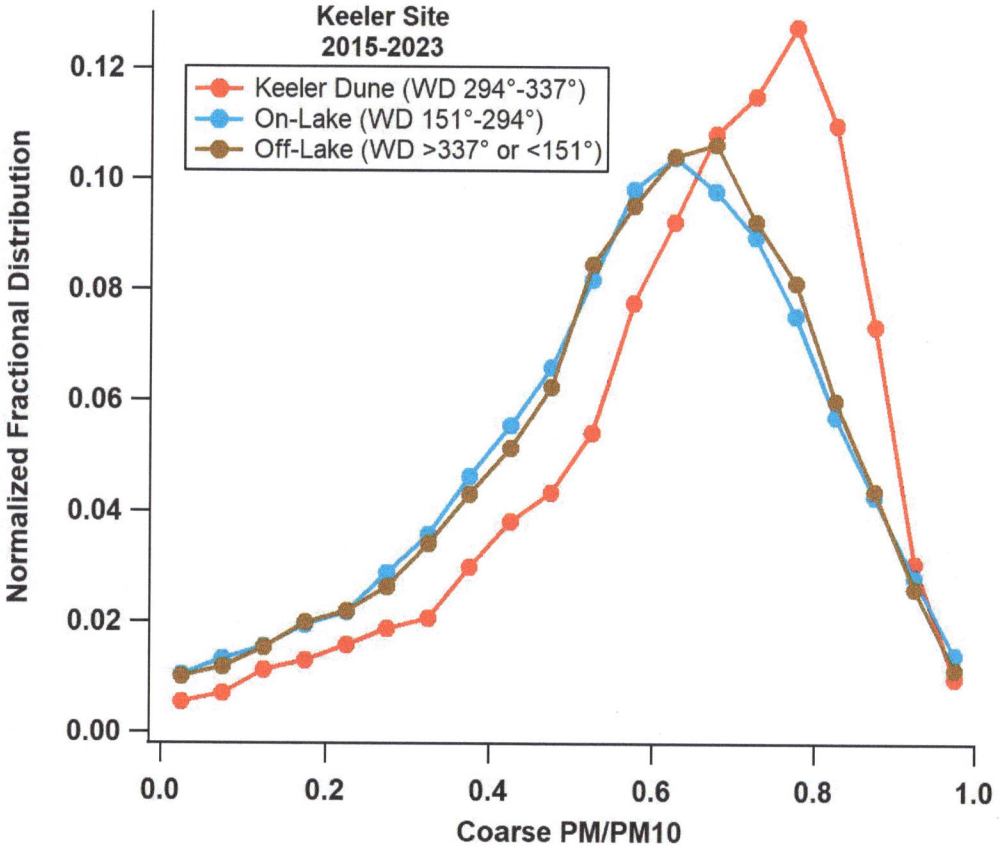

FIGURE 3-20 Normalized distribution of the fraction of PM$_{10}$ coarse particles (i.e., PM in the size range of 2.5–10 μm), measured at Keeler monitoring site. The data are grouped based on wind direction, corresponding to the direction from the Keeler Dunes, on-lake, and other off-lake sources.
SOURCE: Data from California Air Resources Board archive.

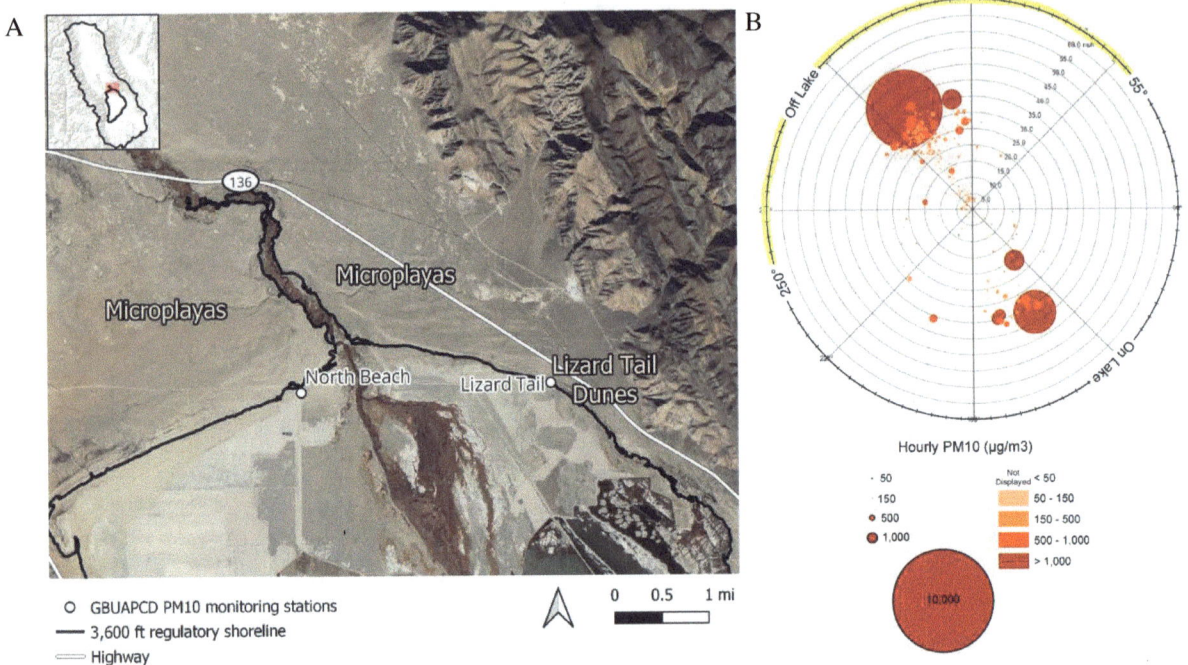

FIGURE 3-21 A) Location of North Beach monitoring site relative to potential local off-lake sources. B) Hourly PM_{10} pollution rose plots at North Beach.
NOTES: A) The term "microplayas" is used here in the manner used by the District. This term is discussed in depth in Chapter 4. B) The markers, sized and colored by the hourly PM_{10} concentrations for data points above 50 $\mu g/m^3$, are mapped out on a polar plot. The angular position on this plot represents the direction from which the wind was blowing at hourly intervals at the site, while the radial distance from the center of the polar plot represents the hourly wind speed. Wildfire smoke events have been removed.
SOURCES: A) Adapted from Google Earth; B) GBUAPCD.

test p-value >0.05). Furthermore, higher concentrations of hourly PM_{10} were observed from the off-lake direction (Figure 3-21B).

Although the exceedance database attributes some off-lake exceedances to human disturbances (e.g., county landfill, roads), other exceedances have different local and regional off-lake sources. All of the regional exceedance days at North Beach from 2017 to 2024 were sourced from a northerly wind event. Although the primary origin of these regional events was determined to be outside the OVPA, the exceedance database noted the potential for additional PM_{10} sources in the valley during some regional events (e.g., "picking up additional sources as the front traveled down the valley" [10/11/2021] and "augmented by local sources north of North Beach between the monitor and the Lone Pine landfill" [05/11/2018]). These exceedances suggest that there may be some occasional, local off-lake sources that lie to the north of North Beach, besides the landfill.

Lone Pine Monitoring Site and Evidence for Other Northern Local Off-Lake Sources

At Lone Pine, a small residential community north of Owens Lake (Figure 3-23A), two to five PM_{10} exceedances occurred in most years between 1993 and 2003 (P. Kidoo, GBUAPCD, personal communication, May 29, 2024). Exceedances in this period were dominated by on-lake sources in some years and by off-lake sources in others. In an analysis of long-term trends using consistent metrics for calculating exceedances (Figure 3-24), there is a statistically significant decreasing trend (Mann-Kendall test p-value <0.01) in the number of

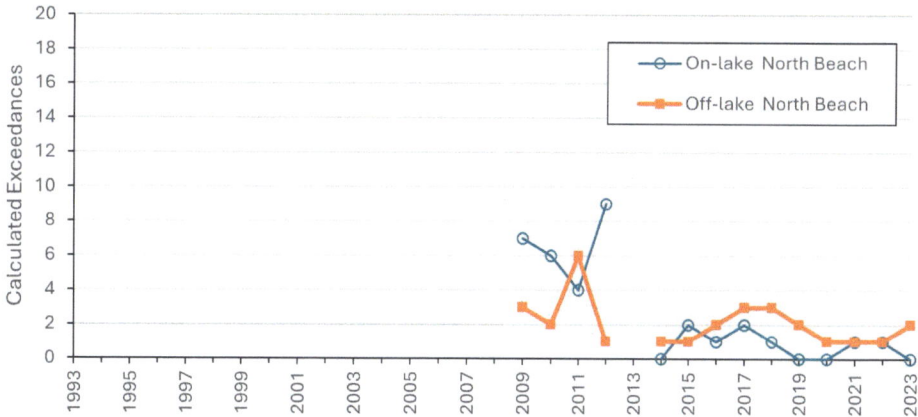

FIGURE 3-22 On-lake and off-lake exceedances at the North Beach monitoring site, with calculations based on consistent metrics over time.
NOTES: These data do not reflect official exceedance counts, but this analysis was conducted to understand long-term trends using consistent metrics. Exceedance counts and on/off-lake attribution based on screened wind directions and a threshold of 150 $\mu g/m^3$ for exceedances. For all years, on/off-lake exceedance counts were purely classified by screened wind directions and may include wildfire smoke events and regional events. For all hours without data when the wind was coming from an opposite direction (on- or off-lake), a background concentration of 20 $\mu g/m^3$ was assumed. Data were not collected between December 2012 and August 2014.
SOURCE: Data from Chris Howard, GBUAPCD, personal communication, November 2024.

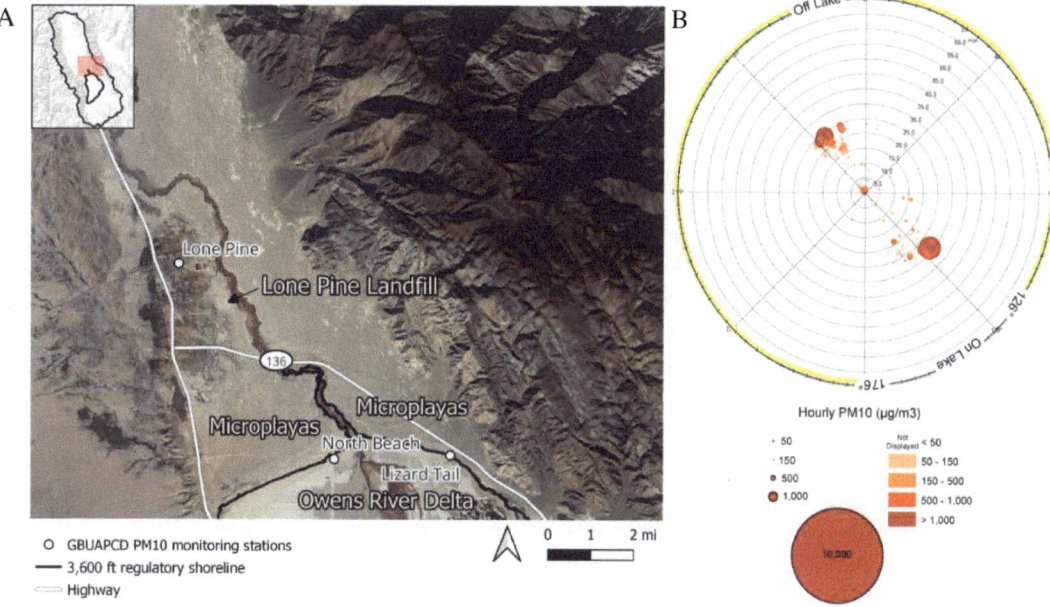

FIGURE 3-23 A) Location of Lone Pine monitoring site relative to potential local off-lake sources. B) Hourly PM_{10} pollution rose plots at Lone Pine.
NOTES: B) The markers, sized and colored by the hourly PM_{10} concentrations for data points above 50 $\mu g/m^3$, are mapped out on a polar plot. The angular position on this plot represents the direction from which the wind was blowing at hourly intervals at the site, while the radial distance from the center of the polar plot represents the hourly wind speed. Wildfire smoke events have been removed.
SOURCE: A) Adapted from Google Earth; B) GBUAPCD.

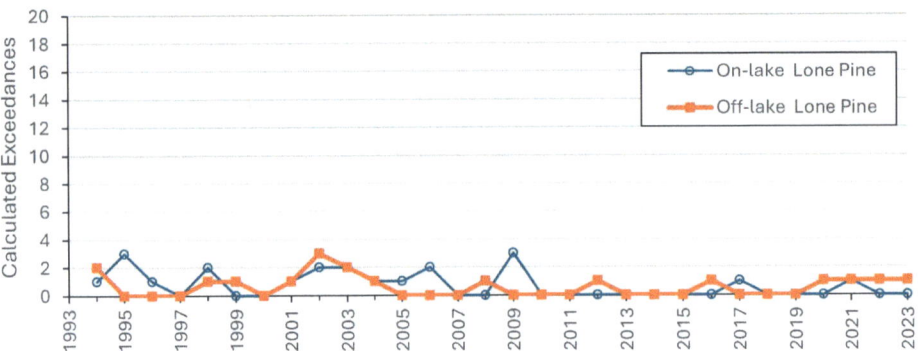

FIGURE 3-24 On-lake and off-lake exceedances at the Lone Pine monitor, with calculations based on consistent metrics over time.
NOTES: These data do not reflect official exceedance counts, but this analysis was conducted to understand long-term trends using consistent metrics. Exceedance counts and on/off-lake attribution based on screened wind directions and a threshold of 150 $\mu g/m^3$ for exceedances. For all years, on/off-lake exceedance counts were purely classified by screened wind directions and may include wildfire smoke events and regional events. For all hours without data when the wind was coming from an opposite direction (on- or off-lake), a background concentration of 20 $\mu g/m^3$ was assumed.
SOURCE: Data from Chris Howard, GBUAPCD, personal communication, November 2024.

calculated PM_{10} exceedances from on-lake sources while there is no significant trend in the number of calculated off-lake exceedances at Lone Pine (Mann-Kendall test p-value >0.05). Based on source attribution in the District's exceedance database, from 2017 to 2024, in addition to regional and wildfire events, there were two exceedances attributed to on-lake sources, one exceedance attributed to off-lake sources (a landfill between Lone Pine and the lake), and one exceedance with mixed sources (Chris Howard, GBUAPCD, personal communication, April 2025).

An additional PM_{10} tapered element oscillating microbalance (TEOM) monitor within the OVPA (Figure 3-25) is located at Fort Independence and is run by the Fort Independence Indian Community of Paiute Indians. There have been 12 daily averages at Fort Independence since 2010 that have been greater than 150 $\mu g/m^3$, all of which have been since 2017. By comparing PM_{10} concentrations at Fort Independence with the monitors on the north side of the lake (Lone Pine, North Beach, and Lizard Tail) and the monitor to the north of the OVPA (Bishop), along with wind direction and information from the exceedance database, the panel estimated that approximately 5 of these 12 events (10/20/2017, 7/20/2022, 1/3/2023, 9/1/2024, and 9/2/2024) could be reasonably attributed to sources local to Fort Independence (Table 3-3). However, there is not enough information to attribute these to a specific source like construction, roads, fallow agricultural fields, open desert, alkali meadow, or other source(s). Of note, the Fort Independence Tribe began several large projects in the fall of 2021, including construction for a Grinding Rock Aggregates facility, construction of several homes, and other developments (Kimberley Mitchell, GBUAPCD, personal communication, March 2025; Sean Scruggs, Fort Independence Indian Reservation, personal communication, January 2025). Thus, while the PM_{10} data suggests that there may be sources near Fort Independence that may be primary causes of some high PM_{10} concentrations observed at the Fort Independence monitor, it is unclear if these sources are derived from active disturbances of the surface (e.g., construction, or highway traffic), or if they are from other off-lake sources of dust that are the focus of this report (Table 1-1).

Overarching Assessment of Local Off-Lake PM_{10} Exceedance Sources

Figure 3-26 summarizes the sources of local off-lake PM_{10} exceedances at all monitoring sites from 2017 to 2024, based on the District's exceedance database (Chris Howard, GBUAPCD, personal communication, April 2025). This analysis suggests that flood deposits, Olancha dunes, and Keeler Dunes are dominant sources of local off-lake exceedances. Human-disturbed surfaces (e.g., landfills, construction sites, roads), alluvial fans,

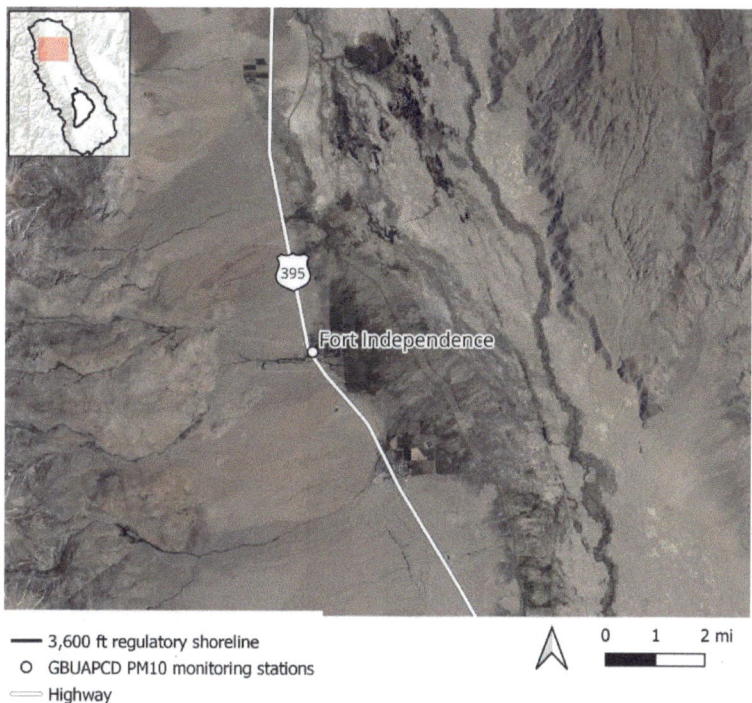

FIGURE 3-25 Location of Fort Independence monitoring site.
SOURCE: Adapted from Google Earth.

TABLE 3-3 Dates Since 2010 with 24-hour Average PM_{10} Concentrations Greater Than 150 $\mu g/m^3$ at the Fort Independence Monitor, Along with Lower Associated 24-hour Average PM_{10} Concentrations at Other Monitors North of the Lake, Indicating a Local Source

Dates	Fort Independence PM_{10} ($\mu g/m^3$)	North Beach PM_{10} ($\mu g/m^3$)	Lone Pine PM_{10} ($\mu g/m^3$)	Lizard Tail PM_{10} ($\mu g/m^3$)	Bishop PM_{10} ($\mu g/m^3$)
10/20/2017	159.2	20.2	18.7	23.7	36.1
7/20/2022	176.0	16.6	14.7	14.1	18.8
1/3/2023	274.8	6.0	7.1	5.4	3.8
9/1/2024	192.8	23.9	21.2	19.6	25.7
9/2/2024	152.9	20.4	19.8	18.3	23.4

NOTES: Bishop monitoring station is located to the north, outside of the OVPA. Note that a power outage across Owens Lake monitoring sites during peak wind speeds on 10/20/17 likely skewed those 24-hour averages.
SOURCE: Data from Chris Howard, GBUAPCD, personal communication, March 2025.

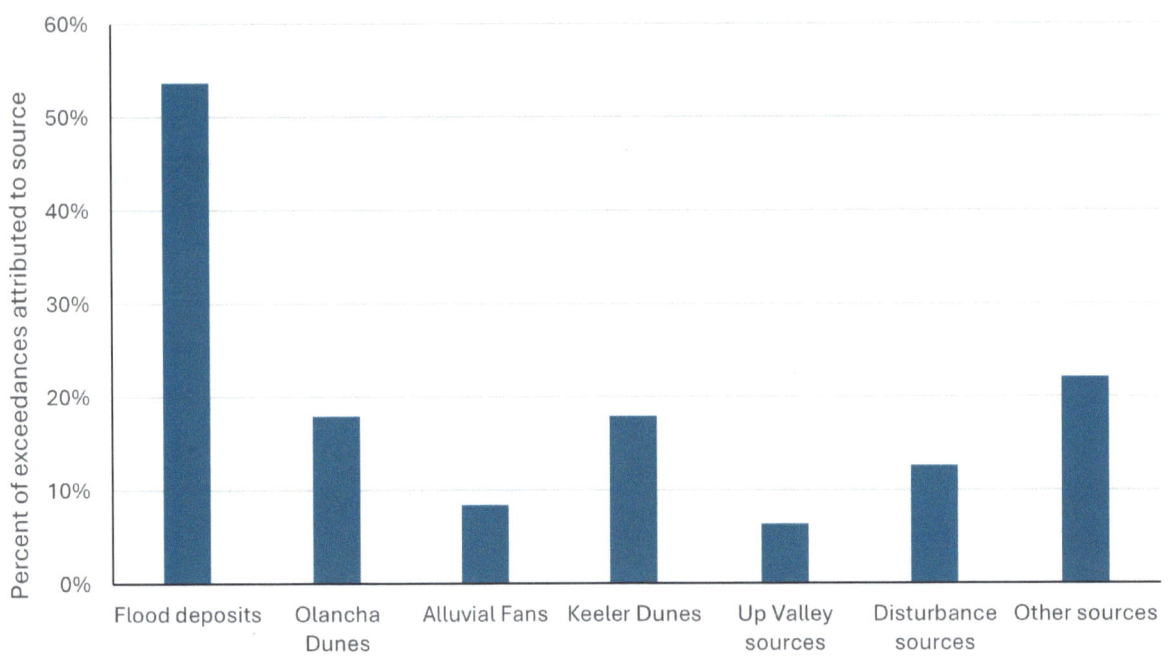

FIGURE 3-26 Summary of sources of local off-lake PM_{10} exceedances at all monitoring sites during 2017–2024.
NOTES: Note that a single exceedance event may be partially attributed to multiple sources. Disturbance sources include roads, landfills, and construction. "Other sources" include attributions to shoreline sources, open areas, sources north of Lizard Tail, sources west of Olancha, Swansea Dunes, and regional sources.
SOURCE: Data from exceedance database, Chris Howard, GBUAPCD, personal communication, August 2024 and April 2025.

and up-valley sources are also important, but individually each contributes to less than 13 percent of documented exceedances. Contribution of "other" sources include open areas, deposits around the shoreline, Swansea Dunes, and regional sources, etc.

TEMPORAL EMISSION TRENDS

An alternative approach to assess the relative role of on- and off-lake sources is to estimate relative emission fluxes and their trends with time. This approach is based on the idea that the product of the concentration and the wind speed measured downwind of an area source is a measure of the emission flux, the emission of PM_{10} per unit area, of the upwind source affecting a monitor. The committee conducted its own analysis of data (methods detailed in Appendix A) that links observed concentrations to emissions accounting for the occurrence of meteorological variations. This approach allows one to compare the impact of on- and off-lake emissions at a given site and examine trends in emission fluxes over time. The empirical relationship between the number of exceedances and emission flux allows us to estimate the expected number of exceedances associated with a given magnitude of emission flux. The results of the modeling are emission fluxes normalized by the mean of the fluxes at Keeler from 2000 to 2023. This normalization isolates the processes that govern fluxes without specifying their uncertain absolute magnitudes.

An analysis of the estimated emissions trends was completed for the monitoring sites that the panel had consistent hourly wind and PM_{10} data since the early 2000s—Dirty Socks and Keeler. Analyses of data from Dirty Socks (Figure 3-27) shows no statistically significant trend for off-lake sources (p-value of 0.96), while estimated on-lake source emissions have had a statistically significant decline over time (p-value <0.01). Keeler (Figure 3-28) has a steep, statistically significant downward trend in on-lake source estimated emissions (p-value <0.01), along

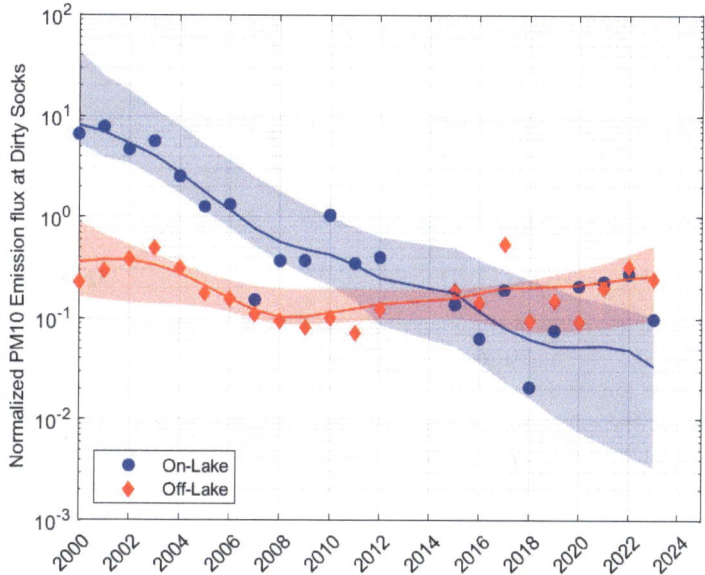

FIGURE 3-27 Trends in off-lake and on-lake normalized PM_{10} emissions at Dirty Socks.
NOTES: The markers are the computed average normalized emission fluxes. The solid lines passing through these points are the mean trends modeled with singular spectrum analysis (SSA). The shaded areas around the trend lines are the 95 percent confidence limits of the trends. Mann-Kendall tests show a statistically significant decreasing trend for on-lake sources (p-value <0.01) and no statistically significant trend for off-lake sources (p-value of 0.96).

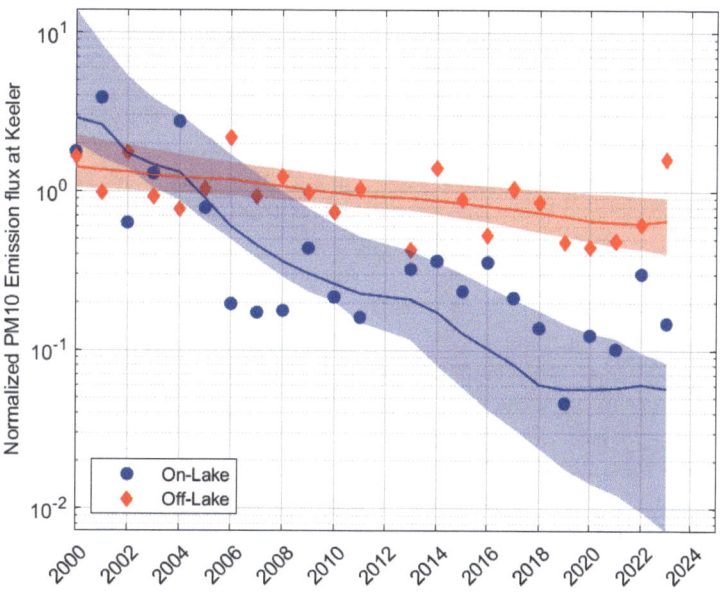

FIGURE 3-28 Trends in off-lake and on-lake normalized emissions at Keeler.
NOTES: The markers are the computed average normalized emission fluxes. The solid lines passing through these points are the mean trends modeled with SSA. The shaded areas around the trend lines are the 95 percent confidence limits of the trends. Mann-Kendall tests show a steep, statistically significant decreasing trend for on-lake sources (p-value <0.01), as well as a more shallow, statistically significant decreasing trend for off-lake sources (p-value<0.01).

with a shallower but statistically significant decreasing trend in off-lake estimated emissions (p-value <0.01), consistent with recent efforts at Keeler Dunes to control off-lake PM_{10} emissions (see Chapter 6 and Box 6-1).

Overall, the analysis indicates that the two monitors with consistent, long-term data show statistically significant declines for on-lake estimated emissions over time, while off-lake emissions either have no trend or are decreasing at a slower rate. At these levels of emission fluxes (see Appendix A, Figure A-1), off-lake emissions will continue to lead to PM_{10} exceedances at monitors in the future if not controlled. Further details are found in Appendix A.

ADDITIONAL MONITORING AND MODELING APPROACHES

Monitoring and modeling are often linked, as air quality models are commonly used to interpret observations or compare modeled predictions with measured observations. There are a number of potentially useful monitoring and modeling approaches that could be conducted to enhance the understanding of which off-lake sources are contributing to continued high PM_{10} levels observed at monitors around or near Owens Lake. This would be most useful for sites that are impacted by emissions from multiple off-lake sources during an exceedance (e.g., Dirty Socks or Keeler). Because current monitors are placed in locations more suited to capture emissions from on-lake sources as opposed to off-lake sources, the panel discussed additional monitoring to better capture and link off-lake emissions to models. Additionally, two general categories of air quality models—receptor vs. source-oriented models—are also discussed in this report.

Additional Imagery

Several cameras around the lake have been useful in the attribution of sources (Figure 3-29), but additional cameras could help distinguish the different sources that contribute to a single exceedance. For example, more cameras southwest of Dirty Socks could distinguish Olancha Dunes emissions from flood deposits emissions. Wider-view cameras north of Lone Pine could better identify sources between Lone Pine and Fort Independence.

Satellite Imagery

Satellite imagery may also be useful in identifying PM_{10} source areas and their evolution over time. Although the OVPA is unusual in that there are a large number of PM_{10} monitoring stations in a relatively small area, they are generally concentrated along the 3,600-ft-elevation regulatory shoreline. In most dust source regions, air quality measurement stations are sparse and located at great distances from each other, leaving vast areas without dust detection or measurement. During the last several decades, remote sensing from orbital platforms have been used to help characterize dust sources. Instrumentation on the satellites measure discrete wavelengths of reflected solar radiation (Ciren, Kondragunta, and Huff 2024) to estimate dust loading in the atmosphere. A limitation is that dust estimates are only available when the satellite is in a position to observe the location, and when the surface is illuminated and without cloud cover. Another limitation of satellite observations is that they are not a direct measure of dust concentrations at the surface but are instead an integrated measure over an atmospheric column.

Some satellites that can provide information on dust are in geosynchronous orbit, such as the Geostationary Operational Environmental Satellites (GOES), the latest version being the GOES-R series. The instruments on these satellites focus on one area of Earth 24 hours a day, but remaining geosynchronous requires an orbit almost 22,236 miles (35,786 km) above Earth, limiting their spatial detail. For most GOES-R wavelengths, the pixel size is about 4 square kilometers. Historically, many estimates of atmospheric dust have been made with the Moderate Resolution Imaging Spectroradiometer (MODIS) instrument package carried on the Terra and Aqua satellites (Baddock et al. 2016; Eibedingil et al. 2024; Kandakji, Gill, and Lee 2020, 2021; Webb and Pierre 2018). Having a non-geosynchronous orbit, Terra and Aqua are in an orbital band of 438 miles (705 km) above the Earth, 1/50th the distance to Earth's surface and thus have greater spatial detail with pixel sizes as small as 0.024 square miles (0.0625 km^2). These estimates of Aerosol Optical Depth (AOD) are comparable to ground-based estimates from the Aerosol Robotic Network (AERONET) at low and middling latitudes, but MODIS-derived AOD estimates are

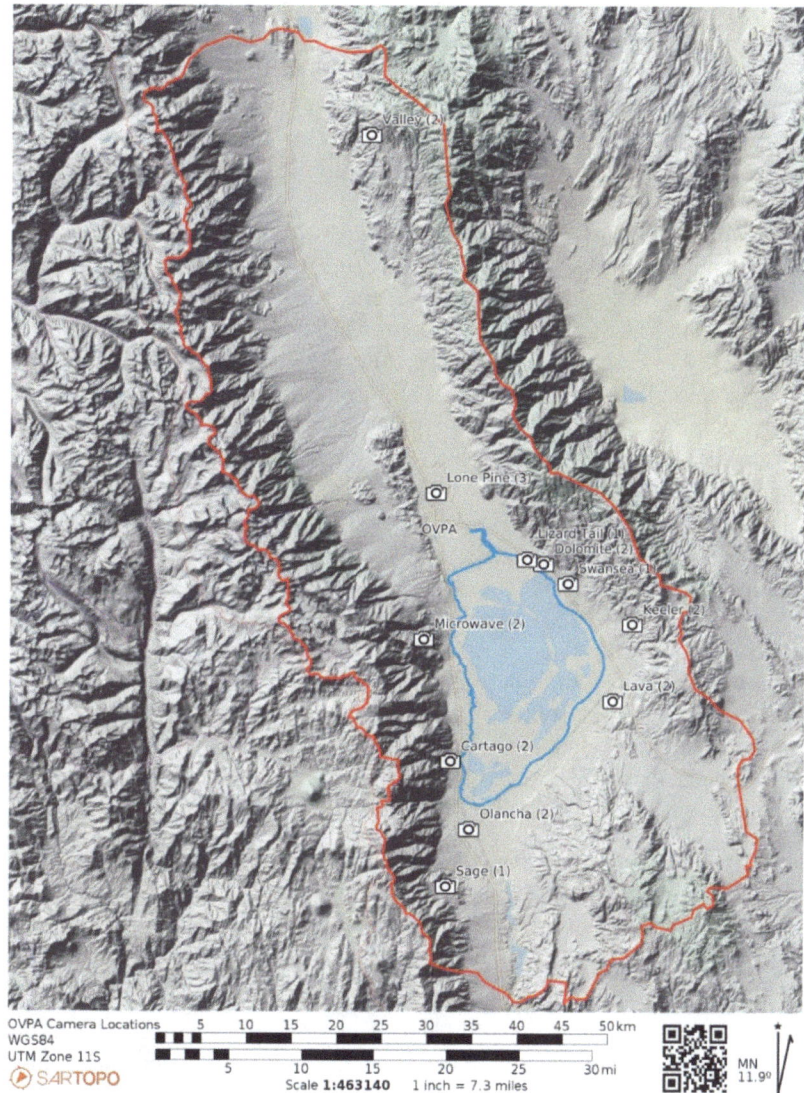

FIGURE 3-29 Sources of dust are assessed by 20 cameras, which are distributed at 11 locations around the lake and in the nearby foothills.
NOTES: Additional cameras also exist outside the OVPA.
SOURCE: Chris Howard, GBUPACD, personal communication, April 2025.

better at high latitudes (Eibedingil et al. 2021). Since the satellites are not geosynchronous, dust outbreaks may be missed when the satellite is not over the dust source.

The MODIS instrument is past its design life and is being replaced by the Visual Infrared Imaging Radiometer Suite (VIIRS) instrument package carried on the Suomi National Polar-Orbiting Partnership spacecraft NOAA-20 and NOAA-21 (NOAA 2024b). VIIRS has a 0.024 square mile (0.0625 km^2) pixel size and 22 wavelength bands (NOAA 2024a), allowing for good dust detection when the instrument is over a cloud-free source region during daylight hours. The National Aeronautics and Space Administration (NASA) is also planning to launch the Multi-Angle Imager for Aerosols (MAIA) instrument in 2026 aboard the PLATiNO-2 satellite (NASA 2024) The PLATiNO-2 completes an orbit every 100 minutes (NASA 2024). As its name suggests, it is specifically aimed at providing more detailed information on aerosols, including an ability to help identify composition, thus providing a linkage to its source (e.g., dust, fire, secondary). MAIA will allow for spectrum analysis of wavelengths bands from 300 to 2,300 nanometers (spanning ultraviolet [UV] to short infrared) and 14 filters (NASA 2024). UV wavelengths detect certain mineral particulates and organic matter, whereas the shorter visible wavelengths detect very small particles according to their diameters, and the short infrared wavelengths are useful for coarse aerosols such as dust and volcanic wash (NASA 2024). MAIA is planned to have a resolution of about 984 ft (300 m; Liu and Diner 2017).

CubeSats are very small (0.3-ft cubes alone or joined in multiples) satellites that orbit at about the same distance from Earth's surface as the satellites carrying the MODIS and VIIRS instrument packages. They are increasing in number and with sufficient numbers flying in formation or constellations, they will be capable of continual coverage of the surface of Earth. High-resolution cameras used in CubeSats have pixel sizes of 10.7 square feet (2.25 m^2) from a 310-mile (500 km) orbit. The Starlink constellations launched by SpaceX are visible satellite constellations that orbit at only 280 miles (450 km) above the surface, 1/80th the altitude of GOES-R. The PlanetScope visual product developed from a constellation of about 130 satellites has been combined with machine learning to estimate ground level $PM_{2.5}$ (not PM_{10}) at a 656-ft (200 m) resolution, with the results suggesting that it can be used to identify areas with elevated PM levels linked to high emissions (Zheng et al. 2020).

NASA's Earth Surface Mineral Dust Source Investigation (EMIT) mission may provide an additional resource for understanding dust sources in the OVPA (NASA n.d.). EMIT, launched in 2022 for deployment on the International Space Station, is a relatively new mission aimed at characterizing the composition of surface dust at a relatively fine resolution (approximately 200 ft or 60 m) and can be teamed with other satellite- and ground-based observations (e.g., from GOES, VIIRS or MAIA in the future) and modeling to characterize OVPA off-lake source regions. Satellite imagery may also play a role in assessing dune movement (Donnallan, Hallet, and Leprince 2015).

The growing information from satellite-based instruments provides a potential source for the District and other stakeholders to identify major dust source areas, and there is little doubt that such capabilities will continue to improve. Furthermore, their use in tandem with ground-based observations and air quality models has proven helpful in characterizing emissions of pollutants (Qu et al. 2022; Wang et al. 2012). A similar approach (e.g., integrating satellite observations of source locations, ground-based observations, and adjoint modeling) could be done to help quantify the contributions of on- and off-lake sources to measured exceedances.

Additional Monitoring

Currently, the OVPA has a relatively large number of PM_{10} monitors given the size of the planning area. However, most of these monitors are located near the historic lake shoreline to capture air masses impacted by on-lake emissions. More detailed identification of off-lake emissions of concern and quantification of their emission strengths can be enhanced by additional monitoring, similar to what was done around Keeler Dunes when a temporary TEOM (a Federal Equivalent Method [FEM] instrument) was installed at the northwestern edge of the Keeler community from March 2015 to February 2020 (Grace Holder and Chris Howard, GBUAPCD, personal communication, January 2025).

While Keeler Dunes has been more intensely studied with near source PM_{10} monitoring and dust emission potential, other off-lake sources such as Olancha Dunes and areas north of the lake are less well characterized. Additional PM_{10} monitoring can help fill that knowledge gap, especially if the observations are linked with air quality modeling analyses. In the last OLSAP report, the panel recommended using lower cost monitoring (NASEM 2020). These lower cost monitors include both very inexpensive monitors (e.g., less than $1,000) and monitors that can be more expensive while still remaining less than the cost of FEM and Federal Reference Method (FRM) monitors to purchase and operate. Lower cost sensors of PM typically run on the principles of PM light scattering, by either nephelometry or optical particle counting techniques (Hagan and Kroll 2020). Studies have found that in dusty areas with significant contribution from coarse particles (i.e., particles between 2.5 and 10 μm), nephelometry-based sensors perform poorly when compared with FRM or FEM measurements of PM_{10} while performance of some of the optical particle counters has been shown to be promising (Alfano et al. 2020; Hagan 2022; Kaur and Kelly 2023; Kuula et al. 2020; Molina Rueda et al. 2023; Ouimette et al. 2022; South Coast Air Quality Management District 2024). For example, during a month-long field study at three sites in Salt Lake Valley, which included sampling during five dust events, optical particle counting-based PM_{10} measurements correlated very well ($r^2 >0.87$) with FEM measurements of PM_{10} by Met-One E-BAM PLUS, with a slope of 0.92–1.39 and root mean squared error (RMSE) of 12–18 $\mu g/m^3$ (Kaur and Kelly 2023). On the other hand, the nephelometry-based data were poorly correlated with E-BAM ($r^2 <0.49$), with RMSE approximately 35–45 $\mu g/m^3$ and slopes of less than 0.099 (Kaur and Kelly 2023). However, in-field calibration of low-performing units improves accuracy and reduces bias to some extent (Alfano et al. 2020; Kaur and Kelly 2023). In many cases, nephelometry-based sensors also perform poorly in high velocity winds, due to the dynamics of the air intake ports. Ideally, air intake ports should be isokinetic and pointed into the wind to draw air at the same velocity as the ambient flow field. These deal situations are difficult to achieve in a turbulent flow field. Another solution that is often employed is to place the sensor in an aspirated chamber where the stilling volume of the chamber permits air intakes that are more closely isokinetic than could be attained in the high velocity flow field. While the data collected by low-cost sensors would not be used to determine compliance with the NAAQS, it can be used to help characterize source areas and for model evaluation, accounting for instrument accuracy. This evidence supports the use of lower cost monitors for characterizing source regions that are less well-studied, particularly if they can be solar powered, rather than using more costly regulatory monitors that would be difficult to use in these conditions (e.g., Chauhan et al. 2022; Riter et al. 2023).

Another type of monitoring that may assist in better defining the distribution of sources are Portable In-Situ Wind Erosion Lab (PI-SWERL) analyses. PI-SWERLs have been used in the OVPA in the past to help quantify PM_{10} emission potential in off-lake areas (Kolesar et al. 2022b). These devices use a rotating annular blade 6 cm from the surface of interest to mimic wind across the surface and create shear stress to generate emissions of PM_{10}. The threshold friction velocity of wind that initiates particulate movement and PM_{10} emissions is then determined (Finnigan 1988; Raupach 1992). Replicate or transect-based testing is necessary because PI-SWERL captures emissions on a small area of the surface.

One advantage of PI-SWERL measurements of surface dust emission potentials is that a range of surface conditions at a single site may be considered. For instance, within a small area of just 2–6 square feet, replicate tests of dust emission potentials from undisturbed or crusted surface conditions may be compared with the dust emission potentials of the same surfaces when disturbed to different extents. Transects could be especially useful in areas like Olancha Dunes to better distinguish the impact of recreational activity on dust emission potentials (Gillies et al. 2022). In addition to disturbance effects, temporal or seasonal factors of dust emission potentials may be measured. Thus, as emissivity varies in response to meteorology, modelling can be used to yield annual predictions based on the expected erosivity of the seasonal winds. As shown in the past applications in the OVPA, linking PI-SWERL measurements, other observations, and modeling results can better quantify the impact of emissions source regions on air quality.

PM_{10} Composition and Receptor Modeling

Having detailed information on the composition of the particulate matter (i.e., the elemental abundances) could also facilitate more detailed source identification. For example, while the Owens Lake bed is a texturally varied mixture of fine clay particles, sodium carbonate, sodium sulfate, and soluble salts (Gill 1995; House, Buck, and Ramelli 2010; Reheis 1997; Tyler et al. 1997), off-lake dune sources are distinctly different in that they are predominately quartz, plagioclase, potassium feldspar with minor amounts of calcite, and other minerals (Lancaster and Bacon 2012; Lancaster et al. 2015). There is evidence of gypsum present in passive dust collectors downwind of the Owens Lake bed (Reheis 1997); however, the source of this gypsum is unknown as it is not present in the surface crusts of playas (Gill 1995). Higher concentrations of heavy metals such as arsenic and antimony are found in Owens Valley alluvium and lake-marginal deposits farther away from the dry bed of the Owens Lake, possibly due to the proximity to naturally occurring minerals in the Inyo Mountains, including the Cerro Gordo Mining District (Reheis 1997; Reheis, Budahn, and Lamothe 2002). Elevated zinc and lead concentrations are also found in dust resuspended from alluvial sediment near Keeler on the northeast side of Owens Lake (Barone et al. 1979; Barone et al. 1981; Cahill et al. 1994). These compositional differences between on- and off-lake sources highlight that chemical compositions could potentially provide some insight into the proportion of on-lake and off-lake dust sources for individual exceedance events. Although it would require high sensitivity, it may also be possible to use the compositional differences between various off-lake sources to better attribute an exceedance event to one or multiple off-lake sources (e.g., contribution from Olancha Dunes vs. flood deposits to exceedances at Dirty Socks).

Elemental analysis of dust from events that are shown to be dominated by different sources can be interpreted using data analysis techniques, often referred to as source apportionment models, discussed below. This will require chemical speciation of the airborne dust, as well as well-defined chemistry of the surrounding source areas (e.g., Frie et al. 2017; 2019; Wang et al. 2023). Chemical speciation of airborne dust can be gained by collecting PM_{10} on filters and then using methods such as X-ray absorption near edge structure (XANES) for metal speciation, inductively coupled plasma mass spectrometry (ICPMS) for elemental measurement, and multi-collector ICPMS analysis for isotopic characterization of the particulate matter on the filters. Filters from the Partisol Sequential Sampler that the District has deployed can be used for such analysis.

The measured composition of PM_{10} (including the elemental composition described above) can be used in receptor models to quantify the contribution of sources with unique chemical signatures to the observed PM_{10}. Receptor models are observation-driven, using the measurements at one or more receptors to estimate how different sources are impacting concentrations at that receptor (Watson 1984). Example receptor models that are commonly used include the Chemical Mass Balance method and the Positive Matrix Factorization or Non-Negative Matrix Factorization (Coulter 2004; Friedlander 1973; Paatero and Tapper 1994; Watson, Cooper, and Huntzicker 1984). Concentrations of individual ions (e.g., sulfate, nitrate, chloride, ammonium, sodium; often measured from filters using ion chromatography), elements (e.g., metals; often measured using x-ray fluorescence or mass spectrometry) and elemental and organic carbon are typical species used in source apportionment analyses. Using the composition of emissions from specific source regions or types, the amount of PM_{10} contributed from the modeled sources (e.g., on-lake, specific dunes, regional transport, and flood deposits) can be quantified. While the Chemical Mass Balance method requires specific knowledge of source composition profiles, Positive Matrix Factorization or Non-Negative Matrix Factorization can simultaneously provide a source composition profile and strength based on its input data of the PM_{10} composition. Ideally, hourly composition data can capture the relatively fast shifts in source areas with changes in wind direction. With enough composition data of PM_{10}, receptor models are generally applied in readily available computer environments, as the models are limited by the availability of the speciated measurements.

Receptor modeling has been conducted at the Oceano Dunes area using a weight of evidence approach to quantitively estimate contributions of sources to exceedances (Wang et al. 2023). This approach involved relationships between combinations of specific PM_{10} species and source emissions. For example, sodium, chloride, magnesium, potassium and sulfate were used to estimate fresh sea salt (FS):

$$FS = fsNa^+ + Cl^- + ssMg^{2+} + ssK^+ + ssCa^{2+} + ssSO_4^{2-}$$

In this example, $fsNa^+$ is the fresh sea salt fraction of sodium ion; Cl^- is the concentration of chloride ion; $ssMg^{2+}$ is the sea salt magnesium; ssK^+ is the sea salt potassium; $ssCa^{2+}$ is the sea salt calcium ion; and $ssSO_4^{2-}$ is the sea salt sulfate ion. Similar relationships were developed for aged sea salt, mineral dust, and other sources. The source relationships developed at Oceano Dunes are unique to the dominant sources surrounding those dunes; however, a similar approach could be applied to Owens Lake once chemical signatures of the sources in Owens Valley are quantified. This approach is very similar to the Chemical Mass Balance method in that it uses known chemical compositions of suspected sources.

Source-Oriented Modeling

Source-oriented models simulate the transport and transformation of pollutants from their emission as they evolve in the atmosphere, typically based on computationally solving the equations governing pollutant dynamics. Those models are based on accurately describing the important physical (e.g., wind velocity, turbulent transport) and chemical processes impacting pollutant concentrations from emissions (which are often uncertain) to their fate. An important use of source-oriented models is to assess our understanding of source emissions strengths to prioritize emission control strategies and better predict emissions and impacts from a specific source on PM_{10} concentrations. If a modeling system performs well under a variety of conditions, that indicates the model is capturing the important sources and processes prevalent in the atmosphere. Lack of agreement, conversely, suggests that there are potentially large gaps in our understanding, which limits the accuracy of the modeling system. For example, if the model results are biased low in comparison to observations, that would suggest that the estimated model source strengths are low or there are missing sources not accounted for in the modeling. If the model simulations are biased high, that would suggest that the source strength estimates are too large. Biases can also be introduced by errors in the meteorological inputs. Diagnosing model errors helps identify errors in our understanding of model processes. Many different source-oriented models have been developed for applications from very local scales (such as the Owens Lake area) to global scales (EPA 2024a; Mejia et al. 2019; Pennington et al. 2024; Vohra et al. 2021).

As described earlier, the District's current modeling approach is to use CALPUFF, along with CALMET as the source of its meteorological inputs to model primarily on-lake source regions. Keeler Dunes is the sole off-lake source currently treated in the District's modeling. With the 2017 revisions to the Guideline on Air Quality Models (Appendix W to 40 C.F.R. § 51), CALPUFF is no longer an Environmental Protection Agency (EPA) preferred model. One concern that arises from the panel's review of the modeling conducted by the District is that the model results can differ substantially from the observations (Figure 3-30). The District developed a hybrid approach using a time-varying background concentration based on monitored off-lake measurements (GBUAPCD 2016), which does improve model performance. However, there are still distinct differences between model results and observations, particularly on days when the observed PM_{10} concentrations are low.

A second concern with the District's current modeling approach is that by only modeling the downwind air quality impacts from on-lake emissions and Keeler Dunes, the ability to quantitatively analyze the impact of other off-lake sources on PM_{10} exceedances and emissions rates is impaired. This information would be valuable for identifying the most effective control program to mitigate future exceedances and in exceptional event analyses. Such information would also reduce bias in the current modeling result, as there are a number of cases where the simulated levels are much less than observed, suggesting that other sources not accounted for in the modeling may be contributing to the exceedances.

Alternative Approaches

One of the most widely used dispersion models in the United States is AERMOD (Chen et al. 2009; Perry et al. 2005). AERMOD is a steady-state plume model and is an EPA-preferred model for dispersion modeling (EPA 2024b). It has a more limited description of atmospheric chemistry than CALPUFF, but this is not a significant issue when modeling PM_{10} mass concentrations. AERMOD is often driven by AERMET, a meteorological preprocessor that utilizes observations to develop meteorological inputs to AERMOD, analogous to CALPUFF being driven by CALMET. AERMOD is computationally fast compared to most three-dimensional models described below or other models that include more complex chemistry. AERMOD has been used to assess source impacts on particulate matter in a variety of applications (Batterman et al. 2014; Chen et al. 2009; Colledge et al. 2015; Jittra et al. 2015; Özkaynak et al. 2013; Perry et al. 2005) including carbon monoxide, nitrogen oxides, particulate matter less than 2.5 µm in diameter, and diesel exhaust emissions, have been associated with adverse human health effects, especially in areas near major roads. In addition to emissions from vehicles, ambient concentrations of air pollutants include contributions from stationary sources and background (or regional.

Other plume models, include the LSPDM (Mejia et al. 2019), LAPMOD (Bellasio et al. 2017; Graff, Strimaitis, and Yamartino 1998), and KSP (Graff, Strimaitis, and Yamartino 1998) space-time varying meteorological conditions, and the desirability of having a model which can yield the probability distribution function (PDF, but it is not apparent if these models have significant advantages over CALPUFF. At Oceano Dunes, the LSPDM is tied to a very fine resolution emissions model based on extensive dust emissions data collected using the PI-SWERL (Etyemezian et al. 2007). At present, there are insufficient PI-SWERL observations in Owens Valley to drive a similar fine-scale emissions estimation approach for the off-lake areas other than Keeler Dunes. However, PI-SWERL testing can be relatively rapid and four replicate tests can be performed in an hour. Thus, it could be possible to obtain the level of spatial detail necessary to model the complex landscapes in Owens Valley at different scales.

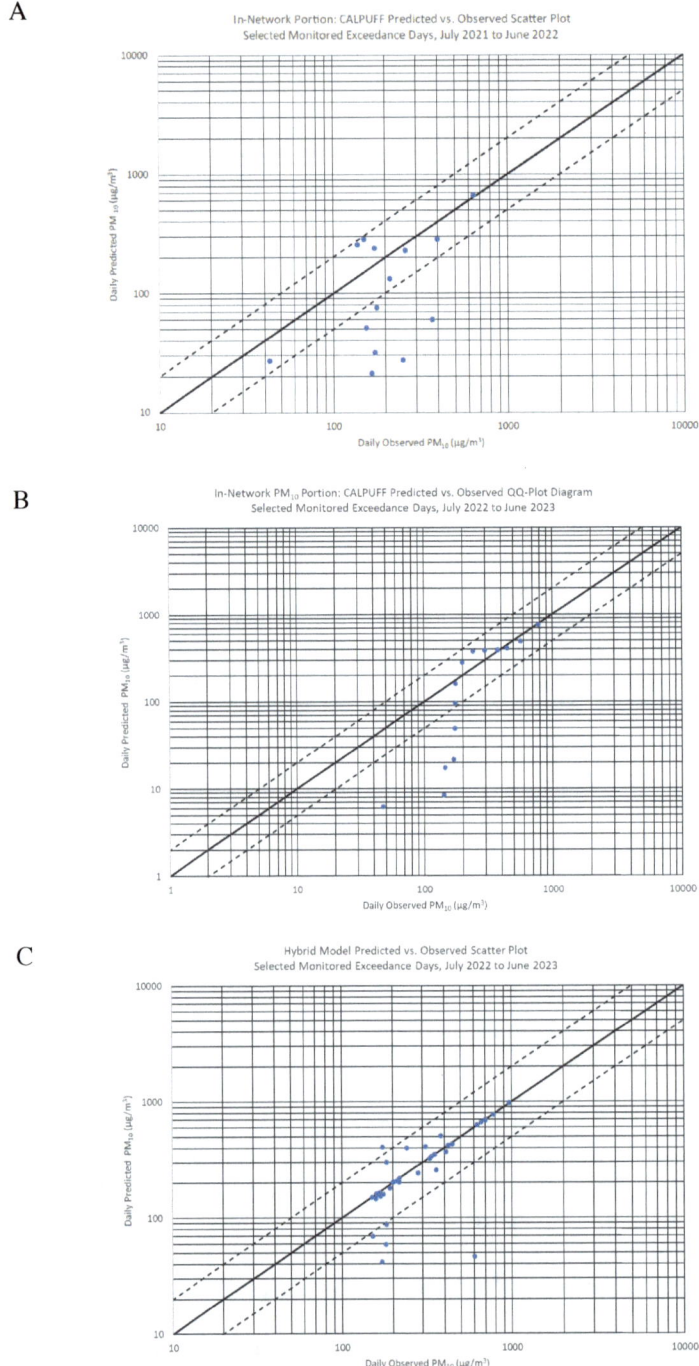

FIGURE 3-30 Simulated CALPUFF model results versus observations for A) 2021–2022 and B) 2022–2023; C) Hybrid model performance for 2022–2023.
NOTES: The scatter between simulated and observed values is also indicative of the variability in emissions, including the variability between the relationships between sand flux, wind speed, and emissions (Gillette, Ono, and Richmond 2004; Kolesar et al. 2022b).
SOURCE: Chris Howard, GBUAPCD, personal communication, September 2024.

The Weather Research and Forecasting (WRF) model is a prognostic meteorological model that solves the basic equations of fluid and energy transport in the atmosphere to provide important fluid-mechanical properties affecting pollutant transport, including wind velocities and turbulent diffusion. CALPUFF can use meteorological outputs from the WRF (Skamarock et al. 2007) and the District's model performance might be improved by including some of these properties. The WRF with chemistry (WRF-Chem) model also uses meteorological parameters from WRF to solve the equations governing pollutant transport from one fixed grid to another, vertically and horizontally (Grell et al. 2005). Contractors for the LADWP used WRF-Chem to conduct preliminary modeling of the region with additional off-lake sources. WRF-Chem is a Eulerian three-dimensional model that differs from plume models in that it is grid-based with a fixed spatial coordinate system. Similar Eulerian models that are widely used include CAMx and the Community Multiscale Air Quality (CMAQ) model.

The use of WRF-Chem or other three-dimensional Eulerian pollutant transport models instead of a plume model has some advantages and disadvantages. Disadvantages include that WRF is computationally more demanding than CALMET, and WRF-Chem (or CAMx and CMAQ) are more computationally demanding than CALPUFF. Another concern is that WRF-generated wind fields can differ substantially from observations (EPA 2019c; Parajuli and Zender 2018). Such differences can alter near-source dispersion calculations, though it is not known how this might manifest in the Owens Lake area. Some advantages include that the Eulerian grid model results are spatially complete, that multiple sources and source areas are treated simultaneously, that the computational time does not increase markedly as new sources are added (although input preparation will), and that the simulations can be used to carry out source-apportionment calculations across the modeling domain, not just along specific trajectories. Treating multiple source areas simultaneously can be particularly important when assessing the cumulative impacts from all the source areas on- and off-lake, though plume models can do so by using plumes from each source area. Some grid-based models have been specifically instrumented to follow emissions from specific sources that can be used to quantify the impacts from on- and off-lake sources at all the monitoring sites in the area (EPA 2025b; Ramboll Americas Engineering Solutions, Inc. 2020). Another advantage of grid-based models is that they can capture the importance of chemical interactions of pollutants from different sources. However, the contributions of atmospheric chemistry to PM_{10} exceedances around Owens Lake is likely small given the short transport times, lack of major local sources of secondary particulate matter formation, and the dominance of direct PM_{10} primary emissions.

Inverse modeling of source-oriented models can be used to better estimate source emissions by adjusting source strengths to better capture observed concentrations. Inverse modeling is typically done by applying a model with the estimated emissions as the input and comparing the model results to the observations, then adjusting the emissions to improve model results. Given multiple observations over multiple times, the estimated emissions are optimized to best capture observations. Zhang (2024) used AERMOD along with multiple linear regression to develop better emissions estimates at a fine scale, as might be done for Owens Valley. More complex approaches include Kalman filtering (Carmichael et al. 2008; Napelenok et al. 2008) and adjoint modeling (Chen et al. 2021; Hakami et al. 2005; Kaiser et al. 2018; Stavrakou and Müller 2006; Zhang et al. 2009).

Value of Including Off-Lake Sources in Models

As noted above, the District includes on-lake source areas and the Keeler Dunes in their CALPUFF modeling. The WRF-Chem model applied by LADWP contractors included some additional off-lake source regions. However, neither appear to consider the broad range of potential off-lake sources contributing to PM_{10} exceedances in the Owens Lake area. Including additional off-lake source areas in air quality analyses of the Owens Lake area would improve the identification and characterization of sources impacting PM_{10} exceedances, provide a platform for air quality planning, and advance assessments of the effectiveness of dust controls.

Historically, the District has used their CALPUFF-based modeling system to demonstrate how controls to on-lake sources and Keeler Dunes will impact air quality (GBUAPCD 2016). This exemplifies a major use of air quality models: to identify effective strategies to meet the region's air quality goals. If additional off-lake controls were to be considered to reduce PM_{10} from off-lake sources, it would be important to include these off-lake areas in the modeling effort. Models can also be used to perform scenario analyses to provide expected source impacts under a range of conditions and controls across the region. Another reason to include a more comprehensive set of off-lake sources in the modeling is to more formally treat the background used in prior CALPUFF modeling. The Hybrid Modeling Approach used by the District includes a time varying background to account for off-lake sources. Inclusion of those sources in the model application would more directly incorporate off-lake source impact on modeled concentrations.

Finally, as discussed in Chapter 5, exceptional event demonstrations are based on thorough analyses of the sources leading to an observed exceedance and an assessment of whether the sources and conditions satisfy the criteria to be considered exceptional events. An important component of an exceptional event demonstration could be using a well-evaluated modeling system to link the observed high levels to specific sources.

CONCLUSIONS AND RECOMMENDATION

Dust control measures have made substantial progress toward reducing the frequency and intensity of on-lake exceedances, but both on-lake and off-lake sources continue to cause PM_{10} exceedances in the OVPA. The relatively consistent number of PM_{10} exceedances that the District has attributed to off-lake sources over the last 25 years, despite trends indicating a declining number of exceedances from on-lake sources, demonstrates the importance of these off-lake sources and suggests that these sources could hinder attainment with the PM_{10} NAAQS in the region.

Conclusion 3-1: Off-lake sources currently contribute the majority of exceedances of the PM_{10} NAAQS at most monitoring sites in the OVPA and are likely to remain important contributors in the future.

Since 2017, the District has used additional information to attribute PM_{10} exceedances to specific sources within the OVPA. This information includes particulate and meteorological data, modeling, cameras, field observations, and media reports of dust storms. These data are compiled into the District's exceedance database. Based on these data, the District classifies each exceedance as one of the following: 1) dust—primarily on-lake sources, 2) dust—primarily local off-lake sources, 3) dust—primarily regional event, 4) wildfire smoke, and 5) mixed—dust and wildfire sources; and provides detailed comments on likely source areas. The panel supports the District's general approach to source apportionment and use of this information to identify a few specific local off-lake sources that cause a disproportionate impact on the PM_{10} exceedances in the OVPA. These sources include flood deposits (including channelized, sheet/overland flow, and impounded flood deposits), Keeler Dunes, Olancha Dunes, alluvial fans, up-valley sources, and anthropogenic disturbances.

Conclusion 3-2: The most frequent local off-lake source of exceedances from 2017 to 2024 is flood deposits, followed by Olancha Dunes and Keeler Dunes.

This District's method for source attribution is useful for assessing broad trends, but the classifications are nonquantitative, and some uncertainties remain in the identification of specific off-lake source areas. For example, the current methodology does not allow for the quantification of PM_{10} contributions from different sources for a single exceedance, and data are often not collected in locations that are ideal to capture detailed information about off-lake sources. Additional measurements and modeling would enable the District to more definitively identify how specific sources have contributed to exceedances and support future air quality management decisions.

Recommendation 3-1: Given the importance of better characterizing contributing sources to individual exceedances from off-lake sources, the Great Basin Unified Air Pollution Control District, the California Air Resource Board (CARB), the U.S. Environmental Protection Agency (EPA) and land owners/ managers should consider supporting the following measurements and modeling:

- **Compositional analysis (species, elements, isotopes) of PM_{10} material to better identify source areas leading or significantly contributing to exceedances.** Compositional data can be obtained from saved

Partisol filters. These compositions can be interpreted with receptor models to better estimate how different sources are impacting concentrations at PM_{10} monitors. Differentiating on- and off-lake sources may be more straightforward, whereas more precision may be required to differentiate between different off-lake sources. Priority for data collection could include those monitors with complex source areas with the largest numbers of exceedances (e.g., Keeler, Dirty Socks).

- **Additional cameras to better attribute specific off-lake sources.** There are a number of cameras around the lake that have been useful in the attribution of sources, but additional cameras could help distinguish the different sources that contribute to a single monitor. For example, more cameras southwest of Dirty Socks could distinguish Olancha Dunes emissions from flood deposits emissions. Additionally, wider-view cameras north of Lone Pine could better identify sources between Lone Pine and Fort Independence.
- **PI-SWERL transects to identify changes to dust emission potential with increased human influence, such as at the Olancha Dunes Off-Highway Vehicle Recreational Area.**
- **Temporary monitoring of PM_{10} to better characterize source areas of concern, potentially including the Olancha Dunes Off-Highway Vehicle Recreational Area, fallow agricultural fields, and former groundwater-dependent vegetation areas to the north of the lake.** The current TEOM monitoring stations around the lake are poorly situated to characterize emissions from specific off-lake areas and quantify the significance of these emissions to existing monitoring stations and communities. There are several potential areas of concern within the OVPA that may benefit from temporary TEOM or low-cost PM_{10} monitors to establish dust emission potential. These measurements can be coordinated with source-oriented modeling activities aimed at source impact quantification. If significant new sources are identified through the use of temporary monitors, then regulatory agencies could consider investing in new (or relocating previously existing) EPA-compliant TEOM stations to better characterize emissions from off-lake areas.
- **PM_{10} modeling of off-lake sources, potentially using AERMOD, an EPA-recommended dispersion model for local-scale modeling, instead of or alongside CALPUFF.** Although there would be additional work to switch from CALPUFF to AERMOD in the near term, the District could benefit from the continued EPA support AERMOD enjoys. Inverse modeling will help quantify on- and off-lake source strengths and how they are evolving in time.

4

Origin and Evolution of Local Off-Lake Sources in Owens Valley

As described in Chapter 3, a number of sources of PM_{10} exist in the Owens Valley Planning Area (OVPA) that are outside the regulatory shoreline of Owens Lake, including flood deposits, Keeler Dunes, Olancha Dunes, sand sheets, alluvial fans, up-valley sources, and a variety of anthropogenic disturbances. In this chapter, the panel evaluates the origin and evolution of these major off-lake sources, or those that might become important sources in the future. Some anthropogenic disturbances that were identified as causing exceedances in Chapter 3 (e.g., landfill, roads, construction) are not expanded upon in this chapter as they are outside the scope of this study (Table 1-1).

There are several methods for determining the origin of emissive landforms and how they developed over time. Radiocarbon dating can provide ages of vegetation, flooding, or evidence of past environments, such as immersion in water or burial of soils. Luminescence dating utilizes sand-sized grains and the property they contain that bleaches to zero as the grain is exposed to sunlight on a transport path. This method can provide the best absolute dating evidence for the timing of when the grain is redeposited in a landform and covered, thus providing information on the history of fine-grained sediment mobility on the landscape. Geochemical and grain size information can also be used to trace sedimentary sources and processes. Finally, aerial imagery and photography can document geomorphic and vegetative changes to landscapes. Geologists and geomorphologists typically use these multiple lines of evidence to infer the origins and evolution of landforms, including the potentially dust-producing ones discussed here.

WINNOWING HYPOTHESIS FOR OFF-LAKE DUST

A process that is potentially relevant for the origin and evolution of a number of PM_{10} sources in the OVPA is "winnowing." As described in Chapter 3, the term "winnowing" refers to the removal of dust from areas above the regulatory shoreline, which had historically originated on the lakebed. This process was described by Ono and Howard (2016) and was included as Appendix G in the 2016 State Implementation Plan (GBUAPCD 2016). Ono and Howard (2016) found that there was a correlation between on- and off-lake exceedances at the Dirty Socks monitor, and that both exceedances were decreasing in frequency and in average concentrations over time. Their analysis implied that the bulk of PM_{10} emission from off-lake sources was originally derived from on-lake sources, and that as PM_{10} emissions from on-lake sources decreased, the derivative PM_{10} in off-lake sources would also decrease. The report states that:

"With the limited supply of sand and dust in these off-lake areas, PM_{10} that is present in the deposited soil is expected to be winnowed out over time, resulting in lower PM_{10} emissions and ambient impacts. Such a decrease in PM_{10}

emissions and impacts were observed at Owens Lake near the Dirty Socks PM$_{10}$ monitor site. A comparison of off-lake and lakebed PM$_{10}$ impacts measured at the Dirty Socks monitor site in this study found that dust from off-lake areas was closely linked to dust activity in adjacent lakebed areas. The results showed that the downward trends in on-lake PM$_{10}$ exceedance numbers and concentration levels closely matched the trends in off-lake areas. A projection of this relationship found that if lakebed source areas cause no new federal exceedances, as expected for the required dust controls on the Owens Lake bed, then the off-lake areas would also show compliance with the federal standard."

More than 8 years later, it appears that the relationship between on- and off-lake dust sources does not hold. Analysis of exceedances from 2001 to 2023 based on consistent metrics indicates that, contrary to the expectations set out in Ono and Howard (2016), the number of off-lake exceedances do not show a significant downward trend over time, even as on-lake exceedances emissions have continued to decrease (Figures 3-6, 3-9, 3-12, 3-18, 3-22, and 3-24). The panel's analysis of estimated emissions at the two sites with long-term data showed similar results for Dirty Socks (Figure 3-27; see also Appendix A). The panel did observe a significant downward trend in estimated off-site PM$_{10}$ emissions at Keeler (Figure 3-28), which is to be expected in response to the Keeler Dunes Dust Control Project (see Box 6-1). Most current dust emissions from off-lake sources are likely not a result of resuspension of material that originated on the lakebed. Instead, the presence and frequent replenishment of highly emissive flood deposits provides ample fine particulates that can be emitted as dust as long as the horizontal flux of sand-sized particles in saltation is sufficient to emit dust-sized particles from the surface. Thus, the panel finds the premise of Ono and Howard (2016) that there is a "limited supply of sand and dust in these off-lake areas" to be false. Given this and the lack of correlation between on- and off-lake exceedances and estimated emissions, the panel finds that winnowing will play a minimal role in reduction in future off-lake PM$_{10}$ exceedances.

NORTHEAST SIDE OF THE LAKE

The northeastern side of Owens Lake is host to several dune fields, sand sheets, and alluvial fan complexes that deliver and rework sediments from the neighboring Inyo Mountains, pre-existing aeolian deposits, former floodplain and delta deposits of the Owens River, and from the Owens Lake bed itself. Some of these landforms, like the Lizard Tail Dunes and Swansea Dunes, have existed on the landscape for long periods of time without substantial change (Lancaster and Bacon 2012; Lancaster et al. 2015). The Lizard Tail and Swansea dunes have not been tied to many exceedances (Chapter 3). In contrast, other landforms to the northeast of the lake, like the Keeler Dunes, have overgone major changes over the last century and are significant contributors to exceedances (Chapter 3).

Keeler Dunes

The Keeler Dune field consists of sand deposits overlying late Holocene alluvial fan deposits that are located below and to the west of the Slate Canyon/Keeler Alluvial Fan Complex (Figure 4-1). The Keeler Dunes and nearby alluvial fan complex are large continuing sources of off-lake exceedances in the vicinity of Owens Lake (Chapter 3). In 2014, the Great Basin Unified Air Pollution Control District (or District) started a project to reduce aeolian transport and PM$_{10}$ emissions on the Keeler Dunes (the Keeler Dunes Dust Control Project), which involved reducing sand transport using artificial roughness and introduction of plant species in an effort to revegetate the dunes (see also Box 6-1).

The District and the Los Angeles Department of Water and Power (LADWP) have provided competing explanations for the activation of the dune field and resultant production of PM$_{10}$. Here, the panel outlines the contextual elements common to both views, details its understanding of the two explanations for activation of the Keeler Dunes, and evaluates the evidence for each.

There is evidence that the Keeler Dunes were a largely stabilized dune field covered mainly by black greasewood (*Sarcobatus vermiculatus*) at the beginning of the 20th century (Figure 4-2). This dune field probably looked very much like the more northerly Swansea and Lizard Tail dunes and like the southern portion of the Keeler Dunes today. Researchers agree that some event or events caused destabilization of portions of the Keeler Dunes (Figure 4-3). Neither the Swansea and Lizard Tail dunes nor the southern portion of the Keeler Dunes appear to

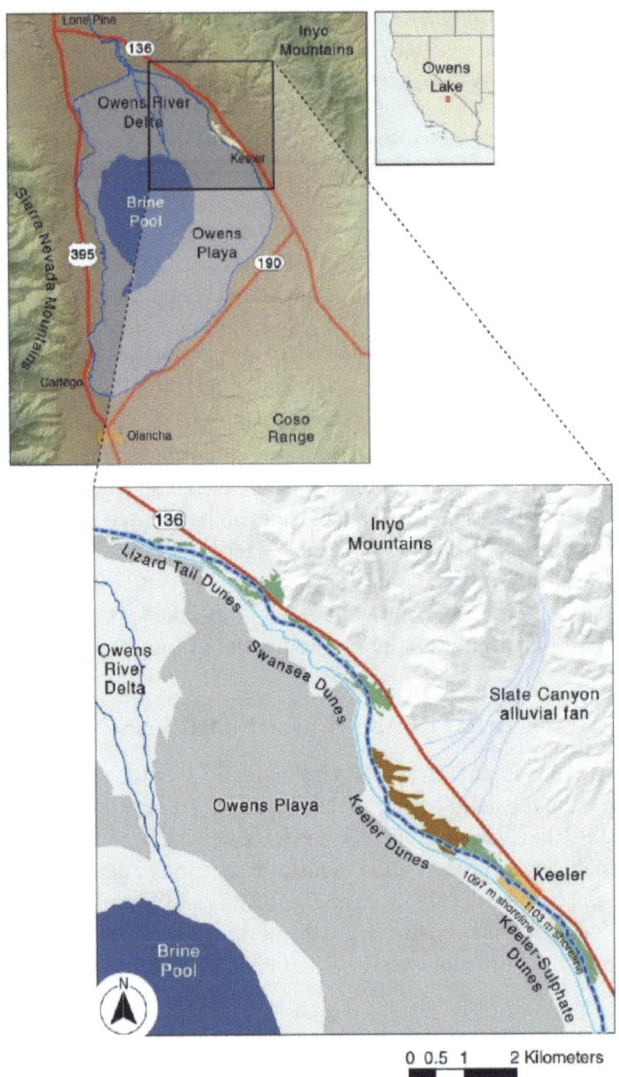

FIGURE 4-1 Dune features on the northwest side of Owens Lake. The active portions of the Keeler Dunes are shown in brown. Inactive portions of the Keeler Dunes and other dune fields are shown in green.
SOURCE: Kolesar et al. (2022b), licensed under CC BY-NC-ND 4.0.

have been impacted in the same way. The main arguments for the remobilization of the Keeler Dunes are referred to below as the "hydrological" and "aeolian" arguments for the sake of simplicity and are outlined briefly here:

1. Hydrological argument: Berms installed in 1954 and 1967 to protect Highway 136 reduced the delivery of surface runoff water to the distal portion of the Keeler/Slate Canyon Fan. As a result, the greasewood that stabilized the dunes was reduced. This line of argument was made primarily in Richards et al. (2022).
2. Aeolian argument: Diversion of water from the lake and subsequent desiccation of the lakebed exposed significant amounts of sand that was blown into the Keeler Dune field. The increase in aeolian transport overwhelmed the vegetation's ability to stabilize the dunes due to feedbacks between the increased sand flux (and abrasion potential) and declining vegetation cover. This line of argument was made primarily in Lancaster and McCarley-Holder (2013).

FIGURE 4-2 A) 1912 and B) 1917 photographs showing greasewood-stabilized dunes in the Keeler area.
NOTES: A) Location 108 at 36.4804 N and -117.8617 E; B) Location 111 at 36.4915 N and -117.8758 E
SOURCE: A) Photographer H. S. Swarth from the Museum of Vertebrate Zoology, University of California (UC) Berkeley; B) Photographer J. Dixon from the Museum of Vertebrate Zoology, UC Berkeley.

FIGURE 4-3 Remains of a greasewood plant, the roots of which have been exposed due to erosion in the Keeler Dunes area.
NOTES: The location of the person's hand is approximately at the top of the root.
SOURCE: Richards et al. (2022), supplementary data, licensed under CC BY-NC-ND 4.0.

These two arguments will be evaluated more below, but first, the common geomorphic and hydrologic context for the Keeler Dunes will be discussed.

Geomorphic Context

Sands from the Keeler Dunes are medium- to fine-grained, moderately to poorly sorted, and have a quartz content of 37–38 percent and a total feldspar content of 40 percent (Lancaster et al. 2015); (Figure 4-4). The composition and mineralogical maturity of the sands from the Owens River and the Keeler Dunes are similar, although the dune sands are slightly more quartz rich, which often occurs in aeolian settings as other minerals (e.g., feldspars) are abraded during aeolian transport, or weathered over time (Lancaster et al. 2015; Muhs 2004). In contrast, sands on the Slate Canyon/Keeler Alluvial Fan Complex have distinctly different mineralogy with higher calcite and lower feldspar content (Figure 4-4A; Lancaster et al. 2015).

Regardless of when sand was deposited (i.e., pre- or post-diversion of water from the lake), mineralogical composition and grain size distributions indicate that the primary source of sand for dune fields in this region appears to be sediment from the Owens River, derived initially from the physical breakdown of granitic rocks of the Sierra Nevada that was then remobilized by wind from the lakebed and exposed river delta into the dune fields during times of low lake levels (Lancaster and Bacon 2012; Lancaster and McCarley-Holder 2013; Lancaster et al. 2015). Past drops in lake levels and periodic floods are known to have increased sand supply to the Owens Lake bed and subsequently to dunes proximal to the lake (Bacon et al. 2018; Lancaster and McCarley-Holder 2013; Lancaster et al. 2015). In recent geological time, optically stimulated luminescence (OSL) ages from the Lizard Tail Dunes indicate two periods of aeolian sand accumulation: 1192–1302 CE and 1592–1712 CE. These periods are also represented in the Keeler Dunes area, suggesting extensive aeolian sand accumulation around 1282–1412 CE and 1563–1712 CE (Bacon et al. 2018; Lancaster and Bacon 2012). The older of these periods closely follows regression from the 3,618 ft (1,103 m) lake high stand that occurred around 1112–1282 CE (Bacon et al. 2018; Lancaster and Bacon 2012), which would have resulted in exposure of lake plain sediments as lake levels lowered. Lancaster and Bacon (2012) provided luminescence ages that indicate aeolian sand was also reactivated during 1930–1976.

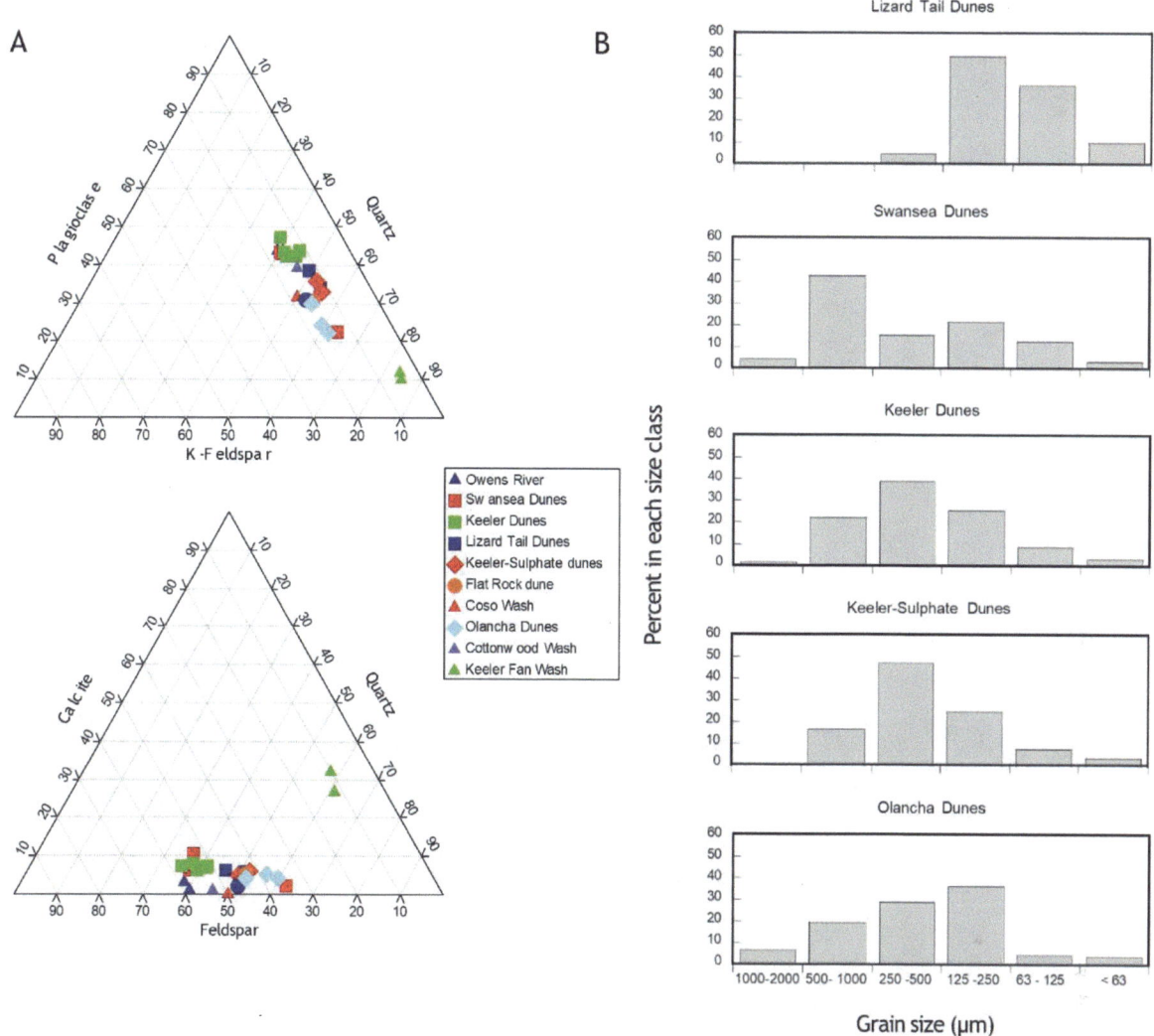

FIGURE 4-4 A) Mineralogical ternary diagrams and B) Particle size distribution for typical sands from major dune fields.
NOTES: Slate Canyon/Keeler Alluvial Fan Complex (labeled in Figure 4-4A as "Keeler Fan Wash") has distinctly different (low feldspar) compositions.
SOURCE: Lancaster et al. (2015).

Flood deposits have a much higher potential for PM_{10} emissions compared to aeolian sands and interfluve[1] fan deposits (Kolesar et al. 2022b). Flood deposits typically occur in distal areas of the fan complex fed by recently active channels and/or within low-lying interdune basins where the dunes locally dam flood flows. In the Keeler Dunes, fine textured (clay–silt) "flood silts" exist within interdune areas as depicted in Figure 4-5 (Lancaster and Bacon 2012) and are generally interbedded with aeolian sands or deposited as a cap on underlying aeolian sands. Periodic flooding is a natural occurrence in desert landscapes that results in mobilization of sediments from slopes, channels, ephemeral washes, and alluvial fans within watersheds. Flood flows also interact with and rework other

---

[1] Interfluve is the "relatively undissected upland between adjacent streams flowing the same general direction" (American Geological Institute 1983).

FIGURE 4-5 Flood silt deposits interbedded with aeolian sands in the Keeler Dunes area.
SOURCE: Lancaster and Bacon (2012).

sedimentary deposits, including aeolian dunes and sand sheets. As such, flood deposits are an episodically renewing part of the natural mosaic of sedimentary features, including at Keeler Dunes.

Hydrologic Context for Vegetation Around Owens Lake

As outlined in chapter 2, there is considerable evidence at present of near-surface groundwater at the margin of the 3,600-ft regulatory shoreline. The wetlands, springs, and seeps that ring the lake (Figure 2-7) and other evidence (Meyers et al. 2021) indicate remarkable stability of groundwater supply in the vicinity of Owens Lake that has not been impacted by the diversion of Owens River, nor by groundwater pumping up-valley. Modeling by Richards et al. (2022) suggests that depth to groundwater in the Keeler Dunes complex is 0–9 m.

Greasewood is a phreatophytic (groundwater-dependent) shrub that stabilizes dunes in the northeastern side of Owens Lake, including around Keeler (Figure 4-2). Studies have noted the capacity of greasewood (Chimner and Cooper 2004; Nichols 1994) to optimize the uptake of water from shallow or deep water sources based on prevailing conditions and to switch to groundwater when vadose zone soil moisture declines (Devitt and Bird 2016). Studies have shown that greasewood can use groundwater to depths of 33–43 ft (10–13 m; Chimner and Cooper 2004; Devitt and Bird 2016; Garcia et al. 2015; Nichols 1994). In the Mono Lake area and the vegetated dune areas at Owens Lake, greasewood has been shown to have roots down at least 10–16 ft (3–5 m) and 15–20 ft (4.5–6 m), respectively (Donovan, Richards, and Muller 1996; Mike Aspinwall, Formation, personal communication, July 30, 2024). Elmore, Mustard, and Manning (2003) reported that the average depth to groundwater in greasewood communities in Owens Valley is 11 ft (3.4 m).

Evaluation of the Hydrological Argument for Keeler Dune Destabilization

Richards et al. (2022) state that "construction of flood control berms (1954, 1967) cut off surface overflow events on the alluvial fan, resulting in subsequent very low plant cover and significant sand movement." Richards et al. (2022) summarizes the evidence in support of the Hydrological Argument for Keeler Dune destabilization, and begins its vegetation analysis in 1944, when the first aerial imagery from the area is available. However, by 1944, considerable time had passed since the final exposure of the nearby lakebed (the North Sand Sheet was exposed between 1917 and 1920, Figure 4-6). Thus, by 1944, sand from the North Sand Sheet (Lancaster and McCarley-Holder 2013) could have already impacted dune stability.

The argument regarding the cause of changes in vegetation cover (Richards et al. 2022) hinges on disruption of periodic overland flow. With hydrologic modeling, Richards et al. (2022) showed that for some years, the amount of surface flow events in areas below the berm would have greatly decreased. Richards et al. (2022) hypothesizes "that interruption of the surface flows by berm construction would have greater impacts on upland, alluvial fan vegetation than on the groundwater-dependent vegetation of the dune complex (Perkins et al. 2018)." This is consistent with other studies that show that shallow-rooted, rainfall-dependent vegetation can be affected from upstream berms (Schlesinger and Jones 1984). The areas immediately below the berms (e.g., regions B3 and C4 in Figure 4-7A), would have been the most affected as these are mostly dominated by precipitation-dependent "upland vegetation" (i.e., shallow-rooted non-phreatophytic vegetation such as most *Atriplex* spp., saltbush, including *A. confertifolia*, *A. parryi*, and *A. hymenelytra*). In fact, a difference in vegetation cover was clear above and below the berm when the panel visited in May 2024 (Figure 4-8). However, Figure 4-9 from Richards et al. (2022), which compares the vegetation cover in 1944 and 1996, does not show large changes in vegetation distribution in the area immediately below the berm in region B3 despite the construction of the berm after 1944. This evidence suggests that there may be other processes that are more strongly controlling vegetation cover. However, in general, the panel finds that Richards et al. (2022) is successful in showing that areas of the Keeler/Slate Canyon Fan that are not groundwater-dependent have higher cover of "upland" (saltbush) vegetation in the absence of highway berms compared to a case in which the berms are present.

In contrast, Richards et al. (2022) does not clearly show that areas that would have had consistent shallow depths to groundwater (e.g., A3 and the lower portions of B3 in Figure 4-7A) were impacted by the berm. Their analysis is consistent with A3 and the lower reaches of B3 having relatively high (approximately 30–40 percent)

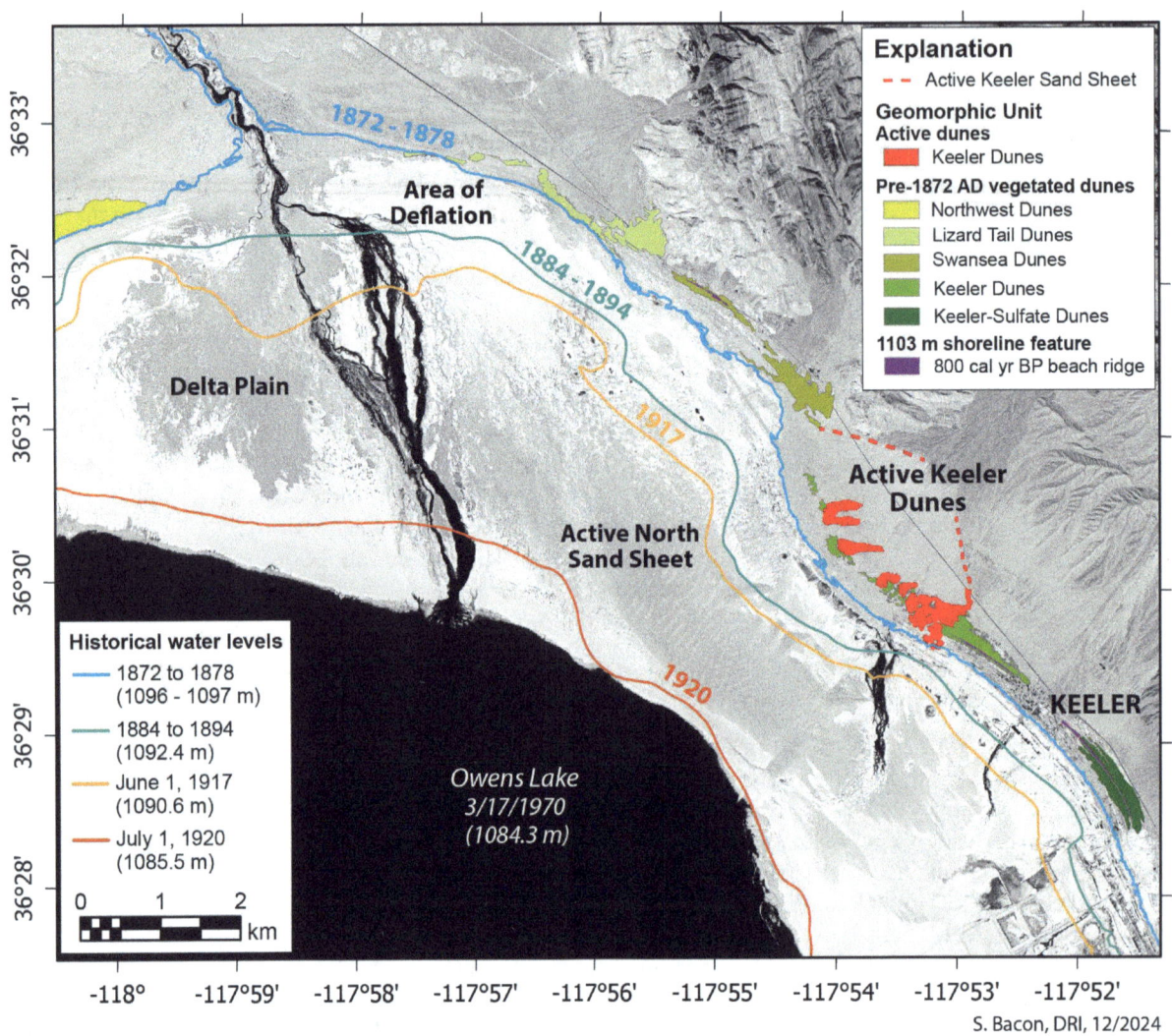

FIGURE 4-6 Reconstruction of lake levels.
SOURCES: Steve Bacon, Desert Research Institution [DRI], personal communication, December 2024, based on data from Lee (1915); Mihevc, Cochran, and Hall (1997); Saint-Armand et al. (1986).

groundwater-dependent vegetation cover given modeled groundwater depths (similar to B4 and B5, see Figure 4-7A) in the absence of any "non-hydrologic" impacts, such as aeolian transport. Indeed, Richards et al. (2022) argues that area A3 (Figure 4-7A) has low vegetation cover compared to other areas with relatively shallow groundwater, possibly due to high sand flux in the area. These impacts to vegetation appear to be present in 1944 and therefore predate any hydrological impacts of the berms themselves. This argument is thus entirely consistent with the view that the northern Keeler Dunes were previously destabilized from some other means (e.g., transport of sands from the North Sand Sheet following water diversion from the lake, per the aeolian argument, see below). Indeed, if disruption of surface hydrology by the berms had impacted dune vegetation, it is logical for this effect to be seen most strongly where the groundwater is deepest and only later witnessed, if ever, in areas with shallow groundwater. In 1944, however, areas with shallow groundwater (A3, particularly in the north) appear to be devoid of vegetation whereas denuded dune vegetation still exists in areas with deeper groundwater (such as in the lower portion of B3; Figure 4-7A). This pattern is thus inconsistent with a hydrologic explanation for vegetation loss.

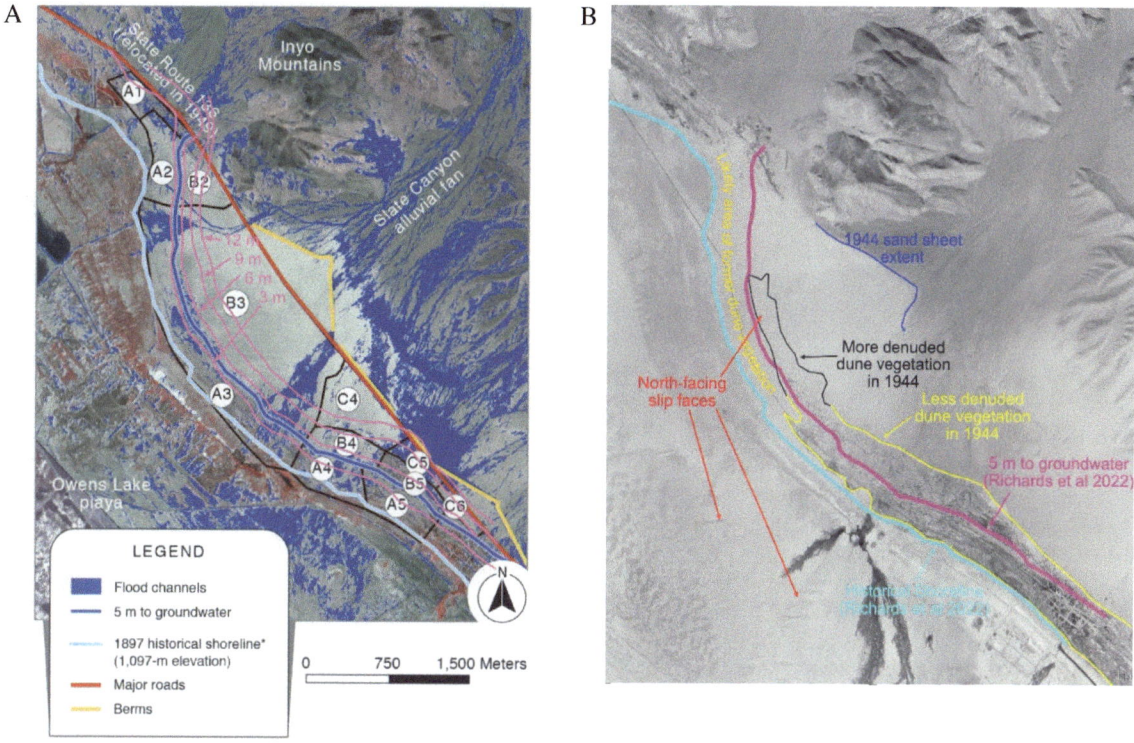

FIGURE 4-7 A) Twelve vegetation provinces as described in Richards et al. (2022) and contours for modeled depth to groundwater, drawn in magenta. B) 1944 aerial photo overlain with the 3,600-ft (1,097 m) historical shoreline and 5 m (16.4 ft) groundwater depth contours.
NOTES: A) Areas B4 and B5 are described in Richards et al. as having groundwater-dependent vegetation despite having modeled depths to groundwater greater than 5 m (16.4 ft; up to approximately 9 m [30 ft]). B) Areas with more denuded dune vegetation (outlined in black) and less denuded dune vegetation (outlined in yellow). Areas between the 5 m (16.4 ft) groundwater contour and the historical shoreline have been labeled as being areas of likely dune (greasewood) vegetation, which is consistent with the modeling results of Richards et al. (2022) in the absence of "non-hydrologic" impacts such as aeolian transport. The apparent extent of the sand sheet in 1944 is also noted in blue. North-facing slip faces indicating transport from on-lake south winds have also been highlighted in red. SIP = State Implementation Plan.
SOURCES: A) Map modified from Richards et al. (2022) with contours from Figure 1 of Richards et al. (2022), licensed under CC BY-NC-ND 4.0; B) 1944 aerial image from LADWP. Contours from Richards et al. (2022) with additional annotations from the panel.

It appears that the modeling approach used by Richards et al. (2022) has little ability to inform the effect of the berms on the ecohydrology of groundwater-dependent vegetation. For example, the modeling approach does not allow the berm to have any effect on the groundwater table, while plant transpiration is prescribed by unchanging root density distributions that depend only upon prescribed vegetation type (upland vs. groundwater-dependent) with root density mainly in the vadose zone. Thus, vegetation growth for groundwater-dependent vegetation, in the model context, always depends on vadose-zone moisture in a strictly prescribed way, even in areas where vegetation has access to groundwater. This approach also does not account for adaptations that vegetation may undergo as a result of changing ecohydrologic conditions, such as root pruning in the vadose zone and increasing water uptake in the saturated zone, perhaps through growth of new roots (e.g., Chimner and Cooper 2004; Nichols 1994). The panel notes that the type of modeling done by Richards et al. (2022) can provide important insights, but the details and limitations of such models are important to consider. Absence of any adaptations to plants in

FIGURE 4-8 Photograph taken May 29, 2024, along Highway 136 berm, facing south from the north end of the berm.
NOTES: The photo shows less cover of upland vegetation (mainly saltbush) downslope of the berm (right) compared to upslope (left). Sand ripples at the right of the image (downslope of the berm and away from the local wind steering caused by the berm in the center of the image) indicate recent eastward (upslope) winds.
SOURCE: Greg Okin, panel member.

response to changing vadose zone water availability or groundwater depth make it impossible to infer from the modeling what impact berm-related changes might have on vegetation in a realistic sense.[2]

Although this study and others show that berms that disrupt overland flow can impact shallow-rooted, rainfall-dependent vegetation (Richards et al. 2022; Schlesinger and Jones 1984), it seems unlikely that disruption of overland flow from the berms in Owens Valley had a major impact on greasewood's ability to access groundwater. Instead, to maintain productivity during dry periods, including after creation of the Highway 136 berms above Keeler Dunes disrupted overland flow, greasewood should be able to access groundwater (or the capillary fringe, which reaches above the groundwater table about 2 ft [0.6 m] for sands and 3 ft [>1 m] for finer soils [Lu and Likos 2004]). Data presented by Mike Aspinwall to the panel on July 20, 2024, indicates that greasewood in the Keeler Dunes area can access and take up groundwater throughout the year (Box 4-1).

Furthermore, greasewood at other sites around Owens Lake do not appear to rely on significant overland flow to maintain productivity. For example, some portions of the Swansea Dunes maintain greasewood cover but do not sit at the distal edge of a large fan like the Slate Canyon/Keeler Fan. In fact, the distance upslope from the Keeler vegetated dune area to the Highway 136 berms is greater than or equal to the distance from the southern portions of the Swansea Dunes to the nearest mountain front, suggesting that greasewood in the area does not rely on significant overland flow to maintain productivity. Evidence from Swansea Dunes, therefore, suggests that greasewood vegetation can be sustained by less moisture from overland flow than is likely received by the Keeler Dunes. Therefore, disruption of overland flow from highway berms above Keeler Dunes would be unlikely to cause the death of mature individuals or cause vegetation collapse.[3]

In summary, the panel evaluated the main paper used to support the hydrological argument for Keeler Dune destabilization (Richards et al. 2022). Although this paper appears to support reduced vegetation cover in upland areas, there is little support for the idea that highway berms led to hydrologic changes that destabilized groundwater-dependent vegetation of the Keeler Dunes. Additional evidence, such as apparent continued utilization of groundwater by greasewood in the area of Keeler Dunes and continued stability of Swansea Dunes, despite limited access to runoff due to its topographic position, further suggests that destabilization of the Keeler Dunes was not caused by hydrologic changes due to installation of highway berms in the 1950s and 1960s.

---

[2] This paragraph was edited after release of the report to clarify the modeling approach used by Richards et al. 2022.
[3] This paragraph was edited after release of the report to clarify area of comparison.

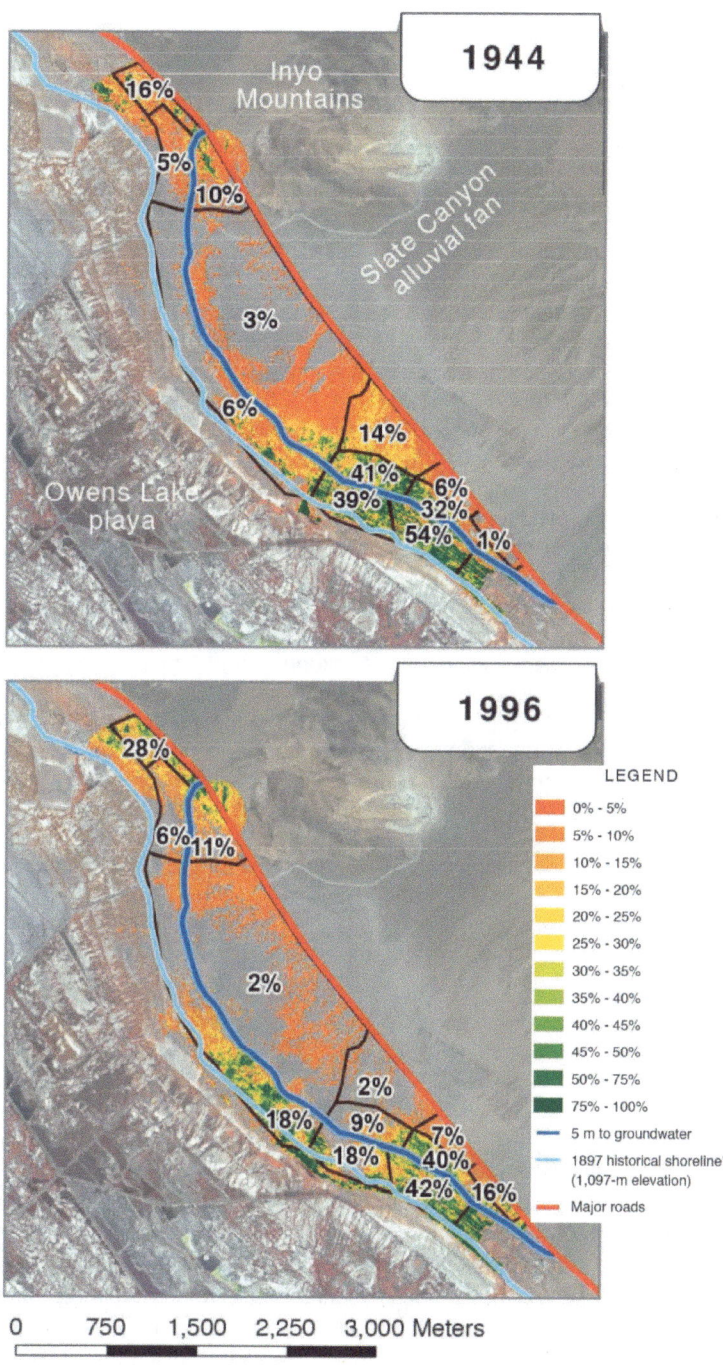

FIGURE 4-9 Plant cover (percentage) for 1944 and 1996 in each of the twelve vegetation provinces in the study area. Berms on the alluvial fan were constructed in 1954 and 1967.
NOTES: Grey areas within the vegetation provinces had non-detectable cover.
SOURCE: Map modified from Richards et al. (2022), licensed under CC BY-NC-ND 4.0.

> **BOX 4-1**
> **Is Greasewood at Keeler Groundwater-Dependent?**
>
> Despite greasewood's dependence on groundwater, shrub cover in vegetated dune areas, which is greater than 90 percent greasewood, does not appear to have a clear relationship with depth to groundwater. Data presented by Mike Aspinwall on July 30, 2024, to the panel show that at sites with a groundwater depth of 15–20 ft (4.5–6 m), shrub cover can be either high (40–50 percent cover) or low (0–10 percent cover). This was used as an argument during presentations to the panel that the dune vegetation uses shallow vadose zone water to support growth when moisture levels are higher. However, this type of dichotomy as seen in the data presented to the panel is also consistent with a bistable state, like the ones proposed by Yizhaq, Ashkenazy, and Tsoar (2007) and Tsoar et al. (2009). Their model showed that vegetation cover hysteresis can arise for a range of wind power conditions. In their model, if there is sufficient precipitation to keep the dunes vegetated, then they will be stabilized with high vegetation cover. However, if disturbed either through reductions in vegetation cover or increases in wind power (drift potential), the dunes can be remobilized in a low vegetation cover state. Although this model uses precipitation instead of soil water/groundwater availability to prescribe water availability to plants, the results should be conceptually transferable to groundwater availability. Thus, a site with differing groundwater availability might have two potential vegetated states (high and low vegetated cover), associated with different dune mobility (low and high), even with constant wind power, due to aeolian feedbacks. In Yizhaq, Ashkenazy, and Tsoar (2009), the vegetation state is dependent on a term related to abrasion. Increasing sand availability (e.g., that caused by the influx of sand from the Northern Sand Sheet, which is below the regulatory shoreline) would be sufficient to increase this term in the Yizhaq, Ashkenazy, and Tsoar (2009) model. Thus, there is a clear mechanistic approach to explain widely divergent vegetation cover, given similar access to groundwater, in the presence of windblown sand-driven mortality.[a]
>
> ---
> [a] This paragraph was edited after release of the report to accurately reflect information provided in the presentation.

Evaluation of the Aeolian Argument for Keeler Dune Destabilization

An alternative explanation for the destabilization of a vegetated Keeler dune field has been offered to the panel. In what we are calling the aeolian argument, desiccation of the lake exposed expansive areas of sand stored in the Owens River delta and North Sand Sheet (an area of the Owens Lake bed immediately south of the Owens River delta; Figure 4-10). Aeolian processes then transported this lakebed material into the area of the Keeler Dunes (particularly the northern portion of the dune field). At the beginning of the 20th century, this portion of the dune field was largely stabilized, and although bare interplant areas may have experienced some aeolian transport, the volume of sand moving through the system was insufficient to cause large-scale mortality of stabilizing vegetation. According to this argument, the addition of sand from the lakebed after exposure of the proximal North Sand Sheet by 1920 (Figure 4-6), however, was sufficient to result in vegetation loss on the formerly stabilized dunes. The initial reduction of plant cover led to increased aeolian activity through the combined effects of reduced sheltering by vegetation as well as increased sediment supply, which generated a positive (amplifying) feedback of vegetation loss and large-scale remobilization of the northern part of the dune field. Lancaster and McCarley-Holder (2013) tracked the evolution of the dune field from 1944 to the beginning of the 21st century, and Lancaster and Bacon (2012) provide additional chronological and stratigraphic context.

Photographic evidence from the 1910s (Figure 4-2) suggests portions of the Keeler Dune field was stabilized by vegetation before diversion of water from the lake exposed considerable sediments available for aeolian

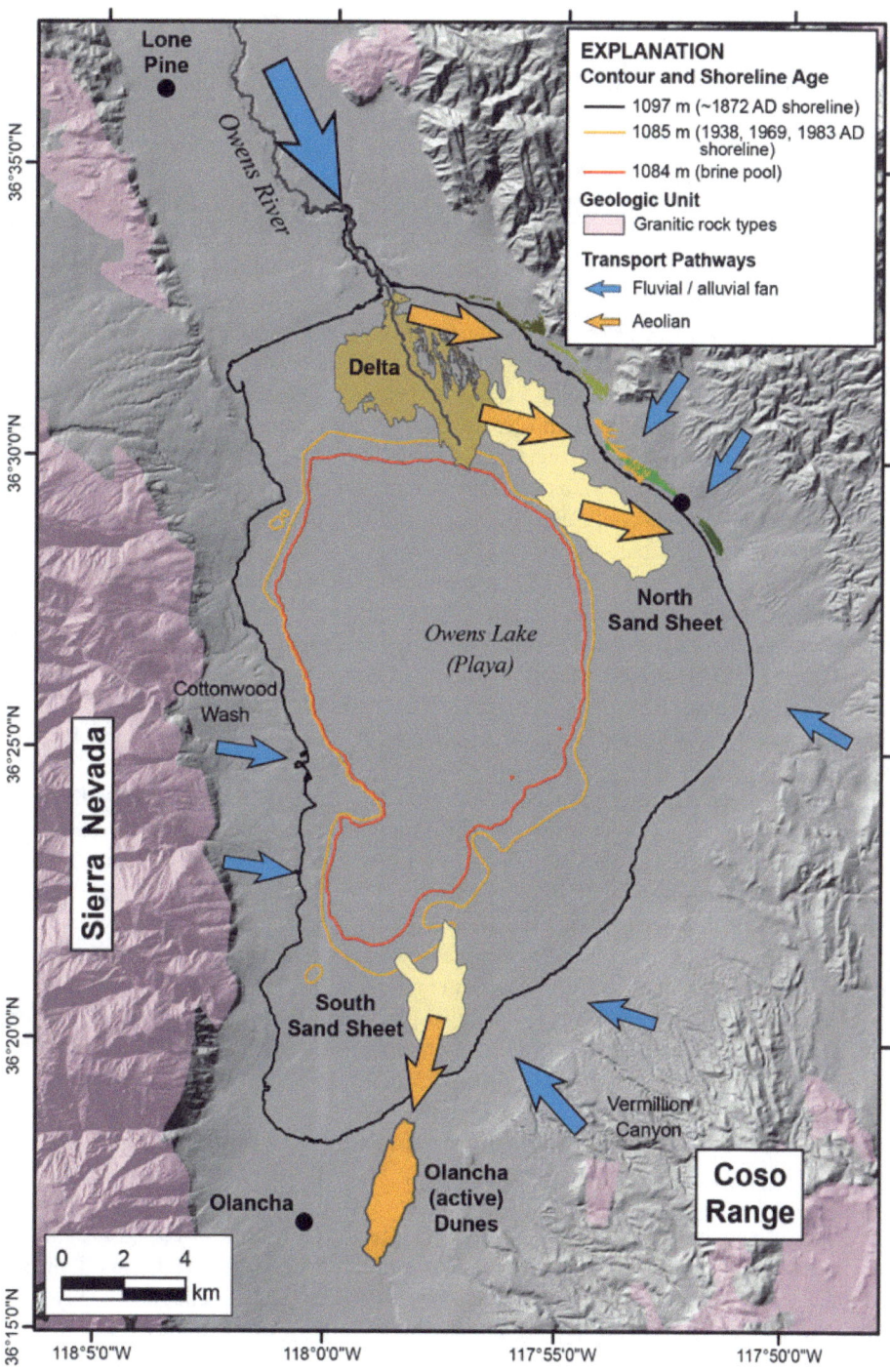

FIGURE 4-10 Schematic of sand transport pathways in the Owens Lake basin showing relationships between fluvial, alluvial, and aeolian transport processes and pathways.
NOTES: The 1097-m shoreline is the 3,600-ft regulatory shoreline. The panel notes that the sand transport pathways indicated for Olancha Dunes may not reflect current wind regimes, which have a northerly resultant sand drift potential (see Figure 4-14).
SOURCE: Lancaster et al. (2015).

transportation. After exposure of the North Sand Sheet on the lakebed, photographic evidence further suggests that there was significant aeolian transport in the Keeler Dunes area, which impacted Highway 136 enough to necessitate installation of sand drift fences. Furthermore, train records indicate that blowing sand and sand drifts were a problem for the railroad in 1954 between milepost 573 and 575, which corresponds approximately to the area of active dunes in the 1944 aerial photo (Figure 4-11; Grace Holder, personal communication, May 2024).

Thus, historical records appear to support the aeolian argument and strongly suggest that the destabilization of the vegetated Keeler Dunes had begun considerably before the building of the protective berms in 1954 and 1967. The OSL age of 1930 and its associated 1 sigma uncertainty range of ± 25 years (Lancaster and Bacon 2012), which is common for relatively young deposits, presents the possibility that the aeolian sand deposit from which the OSL sample was collected was emplaced sometime between 1905 and 1955. This age range overlaps both with the diversion of water and subsequent desiccation of the lakebed as well as the emplacement of the berms in 1954. By itself, this date does not prove destabilization of the Keeler Dunes before 1954, though it is consistent with this view. Without additional sampling, existing OSL data are not conclusive concerning when periods of major dune reworking occurred. Using augers to complete an OSL coring campaign across the sand dunes/sand sheets could provide a more comprehensive set of young ages that are reflective of recent dune activation. While quartz would be the primary target of these studies (as the dominant mineral and best suited to produce higher-resolution OSL ages), feldspar may also provide helpful OSL ages and distinctive luminescence characteristics that allow for increased confidence in the OSL chronology.

Lancaster and McCarley-Holder (2013, p. 281) note that since 1944, when the first aerial imagery exists, "the dunefield has undergone significant changes, including development of well-defined linear and crescentic dunes from an initial small area of partially vegetated dunes, resulting in an increase in the area of the dunes by a factor of 3 since 1944" (Figure 4-12). This has resulted in a relatively large area of active aeolian transport and surface deflation, which has the potential to produce PM_{10} emissions, especially with the presence of interbedded flood silts and aeolian sands within the Keeler Dunes area (Kolesar et al. 2022b; Lancaster and Bacon 2012), which serve as a source for PM_{10} material.

While the southern Keeler Dunes continued to develop and expand from the 1980s through early 2000s, significant wind erosion occurred on the upwind (northern) margins of the dune field (Lancaster and Bacon 2012) with as much as 4 ft (1.2 m) of surface deflation occurring in the interbedded flood silts and aeolian sands. This upwind erosional response corresponds with implementation of dust control measures in 2000, which has served to starve the Keeler Dunes of sand supply and transitioned the system into a negative sediment budget (Lancaster and McCarley-Holder 2013). Meanwhile, the southern region of the Keeler Dunes continued to expand and migrate southeast by reworking existing dune sands. As such, it is apparent that the draining of Owens Lake in the early 20th century consequently increased sand supply to and expansion of the Keeler Dunes, which remain an active dust emissions source. In turn, this has complicated the patterns of fine-grained flood deposits and the extent of their deposition from the Slate Canyon alluvial fan/Keeler Fan Complex within the dunes that provide additional sources of $PM_{10.}$

Several papers have been published that disagree with aspects of this overall narrative. These will be discussed here:

1. Blanton, Kolesar, and Jaffe (2022) have argued that the Keeler/Slate Canyon fan is likely the source of sediment for the Keeler Dunes, at least since construction of the protective berms. However, hydrologic and hydraulic modeling cannot be substituted for mineralogical and geochemical analysis in determining sediment provenance. Such arguments are unconvincing in light of the mineralogical evidence provided by Lancaster et al. (2015; Figure 4-4), which indicate that the mineralogical composition of the dunes is very similar to that of the Owens River and quite dissimilar to the mineralogy of the Keeler/Slate Canyon Fan. More geochemical analyses could further assess this line of evidence.
2. Schmid et al. (2022) argued that dune volume within the dune field was relatively stable over the period of active disturbance (after 1944), which is inconsistent with delivery of large amounts of sand from the Owens Lake bed that could destabilize the Keeler Dunes. The panel has identified several uncertainties in the Schmid et al. (2022) analysis. First, Schmid et al. (2022) only starts its analysis in 1944, at least

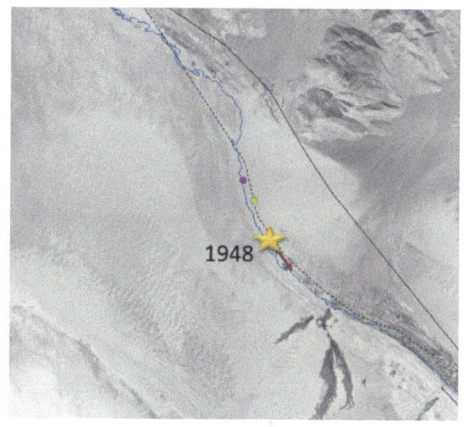

FIGURE 4-11 Historical photos after lake desiccation but before construction of highway berms, showing active dunes associated with significant aeolian transport necessitating installation of sand fences to protect Highway 136 before it was relocated farther up Keeler/Slate Canyon fan.
SOURCES: Photo by Ansel Adams, courtesy of the Ansel Adams Publishing Rights Trust; location map modified from G. Holder, GBUAPCD, personal communication, May 29, 2024.

20 years after the exposure of the nearby lakebed and thus the initiation of transport from the lakebed to the Keeler Dunes. While the panel recognizes this analysis is limited by the availability of the first available aerial imagery, sufficient delivery and storage of sediment in the Keeler Dunes could have occurred before 1944, the effects of which could have been experienced for decades to come. Second, this argument does not consider potential losses of sand from the dune field through transport in the thin sand sheet that is clearly progressing up the Slate Canyon/Keeler Alluvial Fan Complex (Figure 4-13). Even if the volume estimates of Schmid et al. (2022) are accurate, addition of sand from the Owens Lake bed from the west could have been offset by loss of sand in the thin but extensive progressive fan sand sheet. Third, the volume estimates of Schmid et al. (2022) are of insufficient precision to support conclusions about Keeler Dune volumes. The method used to estimate volumes from digital elevation models (DEMs) generated from historical aerial photography in this study is somewhat unconventional. The approach of Schmid et

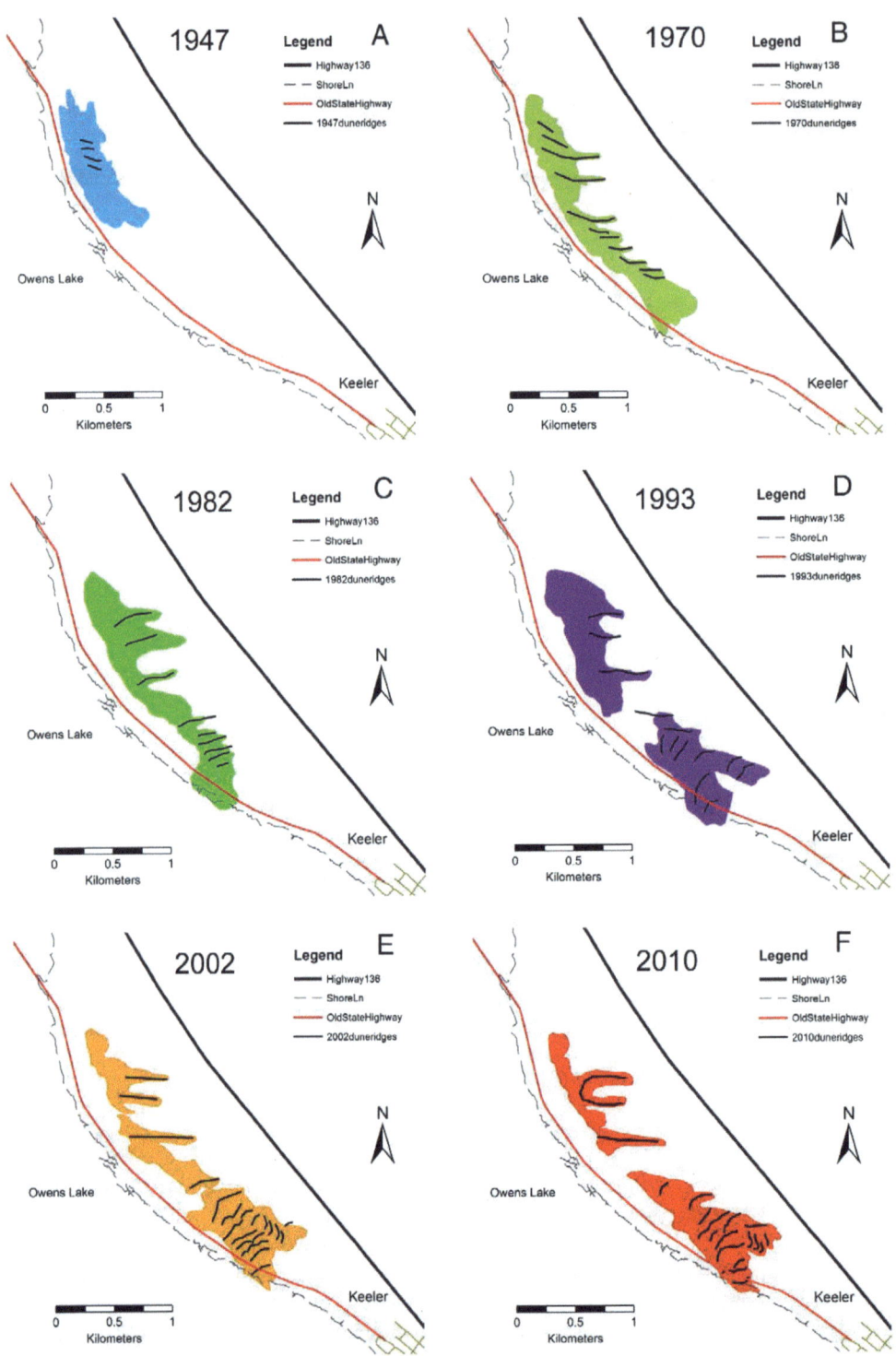

FIGURE 4-12 Footprint of the Keeler Dune field over time indicated by colored regions as identified from historical aerial photography and high-resolution satellite imagery (documented in Lancaster and McCarley-Holder 2013, Table 1). Dark lines within the dune field represent dominant dune crests.
SOURCE: Lancaster and McCarley-Holder (2013).

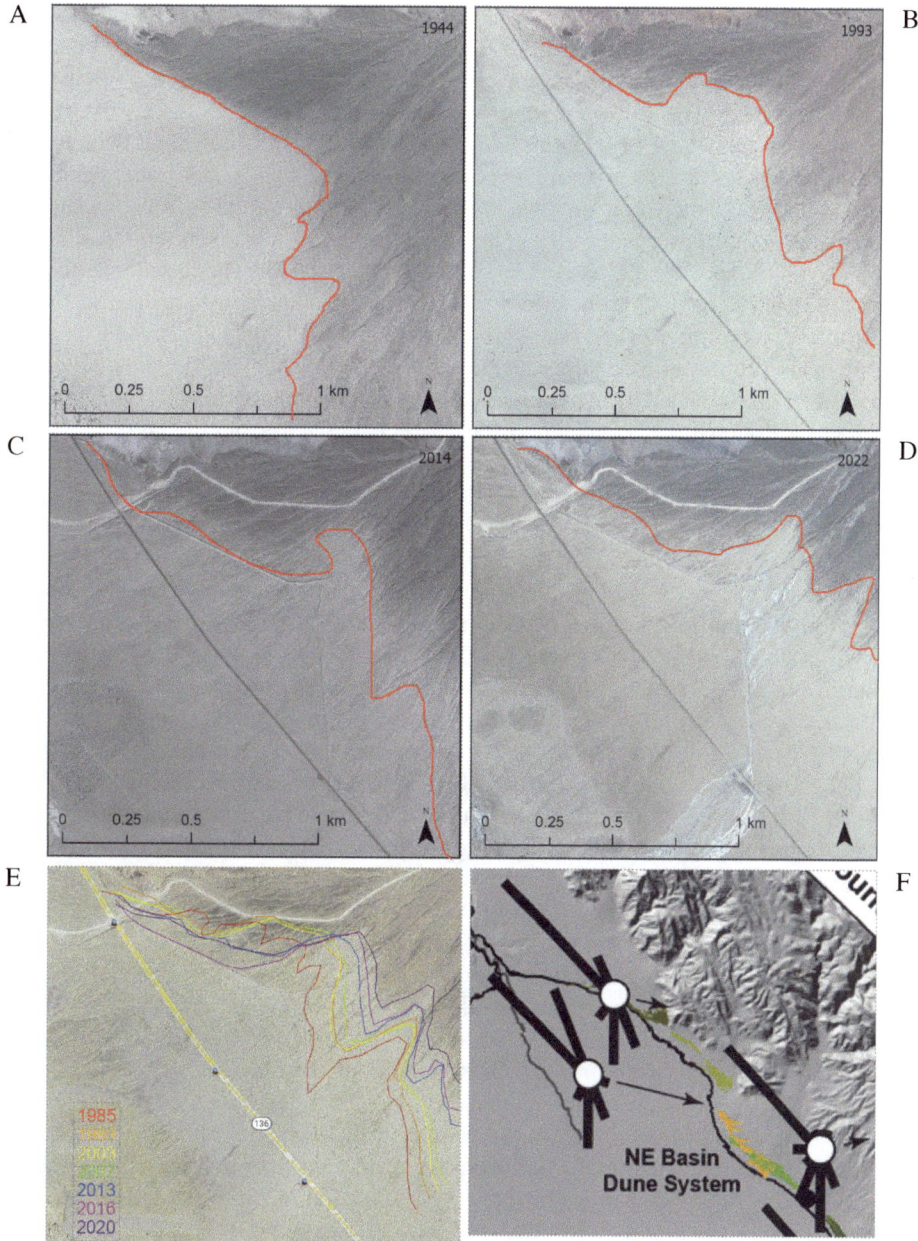

FIGURE 4-13 Historical aerial photography of the Slate Canyon/Keeler Alluvial Fan Complex along Highway 136 approximately 3.2 km northwest of Keeler, CA.
NOTES: Lines on the images delimit the boundary between lighter toned, finer textured aeolian sediments on the left and darker toned, coarser sediments on the surface of the fan complex. Flood diversion structure built in 1954 and 1967 is evident on the northeast side of Highway 136.
SOURCES: A–D) LADWP aerial images, with panel interpretation; E) Google Earth images, with panel interpretation; F) Sand roses have arm lengths proportional to the potential annual sand transportation from different directions. The arrows indicate the direction of the resultant sand drift potential (Lancaster et al. 2015).

al. (2022) is subject to notable uncertainty resulting from the resolution and overlap of historical imagery when used to generate DEMs using traditional photogrammetry. Additional uncertainty is added when defining and interpolating a "sub-dune surface" plain for each photo year, which is used to derive dune heights and raster volumes. More conventional methods use spatial statistics to identify significant elevation changes between successive DEM intervals (Wheaton et al. 2010) and calculate volumetric changes only on locations of significant differences in elevation between the DEMs, not referenced to some arbitrary underlying surface. These more conventional approaches identify multiple error sources, compound them, and remove insignificant raster values (i.e., those that fall below the detection threshold defined by the compounded error budget). Thus, it is not clear that the approach of Schmid et al. (2022) is sufficient for accurate and precise calculation of dune field volumes, and this argument about dune field volumes does not consider additions to the Keeler Dunes prior to the first available aerial imagery (1944) or losses of sand through upslope transport in a thin sand sheet.

3. Kolesar et al. (2022a) estimated a novel index of sand transport concluding that sand transport on the lake occurred in a direction that does not support transport of sediment from the lakebed onto the area of the Keeler Dunes. This study suggests that the wind regime in the area does not support the hypothesis of additional sediment to the Keeler Dunes sourced from the dry lakebed. The study used wind records from a single year (2001) and wind direction measurements from stations on the lake (A Tower and Delta), which are a considerable distance from the historical shoreline in the vicinity of Keeler (approximately 3–8 km). The panel has identified considerable uncertainties associated with this study. First, 1 year of wind data is too minimal to determine decadal-scale patterns in transport direction. Second, as pointed out in Holder et al. (2024), the wind screening conducted by Kolesar et al. (2022a) fails to recognize that "the surface crusting," which is the justification for the screening, "generally occurs on clay and silt dominated soils and is not prevalent on the thick sandy deposits found on most of the northern portion of the bed of Owens Lake (termed the North Sand Sheet) where the analysis was conducted." As a result, the longer records used in the estimates of net sand transport direction (105 degrees, ranging from 17 degrees to 145 degrees) produced by Ono et al. (2011) and Lancaster and McCarley-Holder (2013) are less uncertain (Figure 4-14). These directions support transport from the North Sand Sheet to the area of the Keeler Dunes (Figure 4-14). Third, it is also important to note that resultant drift potential or net transport directions, when calculated from station data, do not fully represent sediment dynamics in areas with complex topography or variable vegetation. On a smooth, unvegetated surface, winds from various directions may move a sand grain in the direction of net transport over a period of time. However, in complex terrain or areas with variable vegetation cover, sediment cannot be expected to experience the same transport patterns (Figures 4-15 and 4-16). Use of winds at a single point for calculation of resultant sand drift potential (Kolesar et al. 2022b; Lancaster and McCarley-Holder 2013) will, therefore, not capture actual sediment movement over areas with complex topography or variable vegetation cover. Indeed, the arrangement of topography in the vicinity of Keeler Dunes is likely to steer transport north of the generally east/southeast resultant drift potentials predicted from winds measured on the lakebed (Figure 4-15). And because vegetation above the 3,600-ft shoreline protects sand blown into it from being blown back out again (per Figure 4-16), transport off the lake could be increased farther still from what might be expected from resultant sand drift potential estimated from winds measured on the lakebed. Finally, the analysis of Kolesar et al. (2022b) and consequent arguments do not account for a wealth of geomorphic information that is probably more relevant for understanding decadal-scale transport than a single year's data. For instance, the 1944 aerial photos show north-facing slip faces on relatively large dunes on the lakebed (Figure 4-7) indicating significant periods of dominantly northward wind transport near the 3,600-ft contour, which likely would have been able to transport sediment off the lake in the area of the Keeler Dunes. Once transported into the Keeler Dunes, microtopography of the dunes and/or dune vegetation could prevent transport from opposing winds back onto the lakebed. In addition, images of the area tend to corroborate the more northward aeolian sand drift potential roses of Lancaster et al. (2015) compared to those of Kolesar et al. (2022b). Figure 4-13 shows lighter toned, finer textured aeolian sediments on the fan complex. The morphology and orientation of these deposits support transport in an

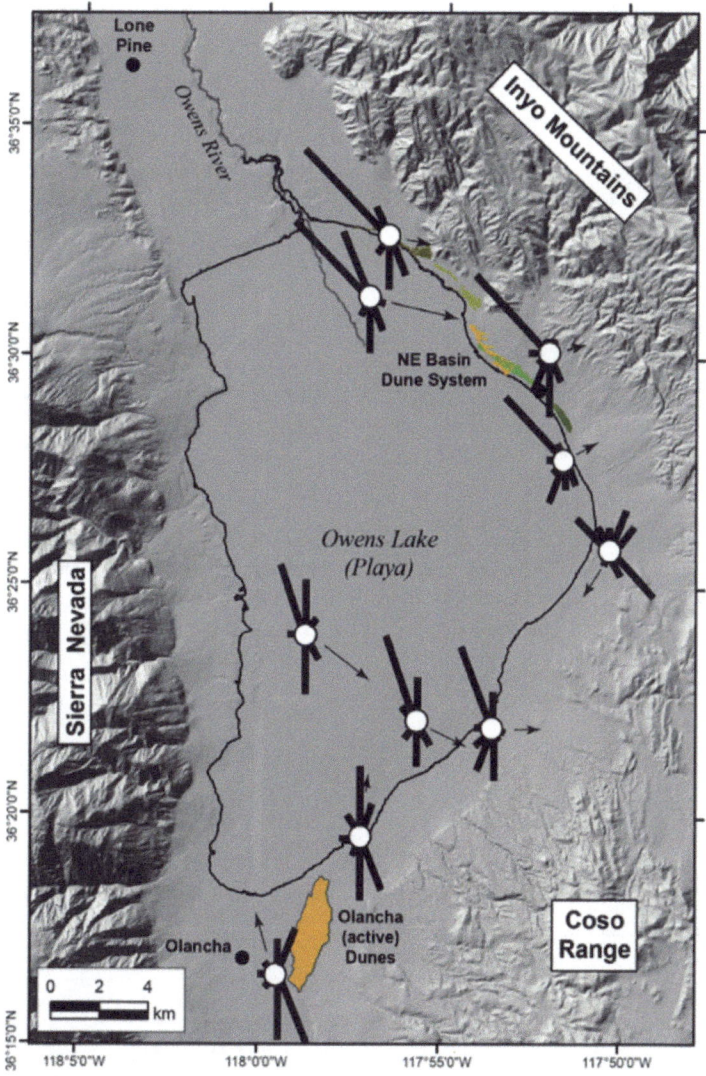

FIGURE 4-14 Sand roses.
NOTES: Arm length is proportional to the annual sand transportation from different directions. The arrows indicate the direction of the resultant sand drift potential.
SOURCE: Lancaster et al. (2015).

upslope direction from the general direction of the lakebed as suggested by the resultant sand drift potential vector. The darker toned surfaces to the right of the line in Figure 4-13 reflect coarser sediments of the alluvial fan/wash complex. The boundary between aeolian and alluvial surficial sediments is dynamic and has generally progressed up slope in response to continued sediment supply from the dunes. This movement is evident by comparing the position of the boundary between 1944 and 1993. The migration continues to present day (Figure 4-13D), despite implementation of dust mitigation efforts on the lakebed in the early 2000s. Small dunes and aeolian sand sheets are also found upwind (southwest) of, and migrating over, the flood diversion berms built in 1954 and 1967 on the northeast side of Highway 136 (Holder et al. 2024). The boundary between lighter- and darker-toned sediments appears to be influenced

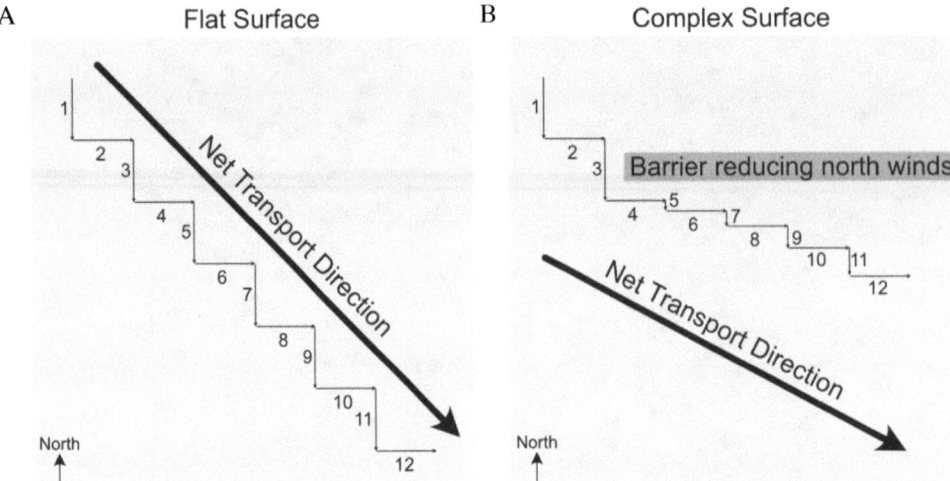

FIGURE 4-15 A) A depiction on a flat surface of particle transport for 12 alternating wind events that move the particle southward and eastward, parallel to the winds, resulting in a net southeast transport direction. B) A depiction of what might happen with the same 12 winds if a topographic barrier reduced the capacity of the north winds to transport particles in its lee toward the south. Events 5, 7, 9, and 11 in the case with the barrier result in less southward transport compared to the flat case, and thus the resultant direction of wind is east/southeast rather than southeast.

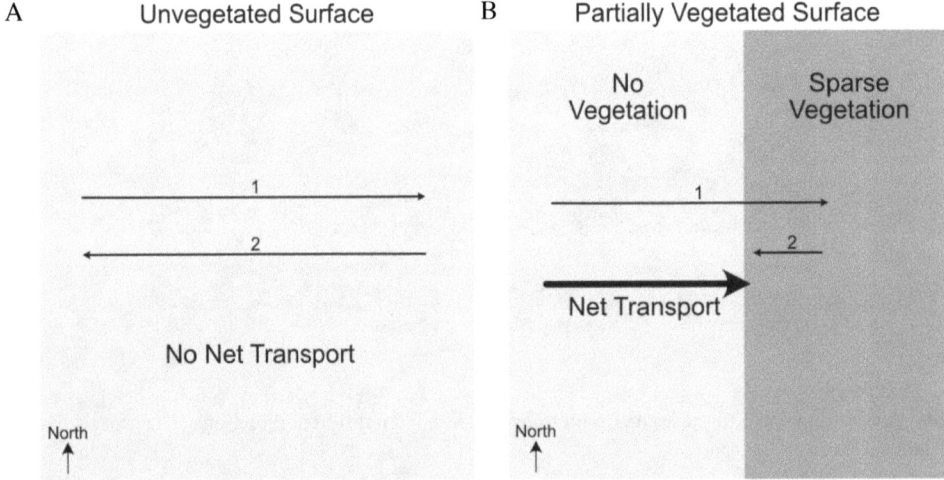

FIGURE 4-16 A) A depiction on a flat unvegetated surface of particle transport for two opposing wind events that move the particle eastward, then westward, parallel to the winds, resulting in no net particle transport. B) A depiction of what might happen with the same two winds if sand was blown into an area of sparse vegetation during the first wind event. Reduced capacity of wind to transport particles in sparse vegetation due to the vegetation's sheltering effect would result in less westward transport during the second event and net sand transport toward the vegetated area.

by periodic floods on the fan complex that transport some of the aeolian sands back toward the dunes and lower valley floor, as evident by comparison of the 1993 to 2014 images (Figure 4-13B, C) or the image that overlays data from 1985–2020 (Figure 4-13E). This structure now diverts floodwaters and fine sediments from the fan complex to two focal locations approximately 1.3 km apart along Highway 136. These locations require regular maintenance following flood events to clear the highway and drainage culverts of sediments, which include aeolian sands that continue to migrate back upslope from the Keeler Dunes and lower elevation areas.

Panel Assessment of the Origin and Evolution of Keeler Dunes

There is considerable historical evidence that supports the aeolian argument for the destabilization and growth of the Keeler Dunes. The counterarguments offered by Blanton, Kolesar, and Jaffe (2022), Kolesar et al. (2022a), and Schmid et al. (2022), do not present convincing refutations of the basic processes underlying the aeolian argument. In contrast, the panel found the hydrological argument to be unconvincing and inconsistent with other observational evidence. The panel finds the aeolian argument to be the most likely explanation for the current active Keeler dune field and support the view that desiccation of Owens Lake is ultimately responsible for the destabilization of the historically vegetated Keeler Dunes.

With the destabilization of the vegetated Keeler Dunes and its emergence as an active dune field, there is clearly sufficient saltation to drive PM_{10} emissions from the flood silts in the area (as seen by the continued exceedances). Because fine-grained material for PM_{10} emissions from Keeler Dunes are continually replenished via flood deposits, there is no reason to believe that PM_{10} exceedances from the Keeler Dunes will stop as on-lake sediment sources are controlled. However, even though the PM_{10} emitted from the Keeler Dunes may not have originated on the lakebed, the desiccation of the lake still increased dust emission from the Keeler Dunes by destabilizing the vegetated dunes at Keeler and increasing the horizontal flux of saltating material that, through sandblasting, emits PM_{10} from the flood deposits. If the vegetated dunes had not been destabilized by excess sand from the dry lakebed, the current and past levels of dust emissions at Keeler Dunes would not have occurred.

Berm-Related Channel/Flood Deposits at Keeler Dunes

Several constructed berms northeast of Highway 136 intentionally modified the surface hydrology of overland flow to focus floodwaters to specific points of discharge along the highways. The panel did not examine each berm-related flood deposit individually but instead focused only on the berms installed in 1954 and 1967 on the Keeler/Slate Canyon Fan. As discussed above, the panel did not find evidence that the influence of the berms on the distal portion of the fan was sufficient to have modified the morphodynamics of the Keeler Dunes (i.e., per the hydrological argument). It is evident, however, that the berm has recently had appreciable, localized impacts on the distribution of flood deposits in the Keeler Dunes region. For instance, the 2022 aerial photograph in Figure 4-13D shows pronounced channel incision and flood channel deposits below the berms following impacts from the remnants of Hurricane Kay in September of 2022. These types of events and their related deposits appear to be infrequent as they are not evident in other aerial photographs. However, these infrequent events could still contribute appreciably to source materials for PM_{10} emissions detected at the Keeler monitoring station. Further investigation is needed to determine how the changes in the distribution of the flood deposits due to the berms at this and other locations have changed dust emission potential.

SOUTHERN SIDE OF THE LAKE

The southern side of Owens Lake is host to another major dune field, sand sheets, and multiple alluvial channel/wash systems that deliver and rework sediments from the neighboring Coso Range. As discussed in Chapter 3, several of these features have been identified by the District as contributing sources to exceedances.

Centennial and Coso Washes

The Centennial Wash drains the northern portion of the Coso Range ending in a large alluvial fan that extends into the Owens Lake bed, and the Coso Wash similarly terminates in an alluvial fan located to the southwest of the Centennial Wash (Figure 3-11). As with all such desert washes, the Centennial and Coso wash/fan complexes consist of a series of channels set into the fan, filled with generally fine-grained (compared to the fan), wind-erodible sediment, and kept free of vegetation through occasional flows of water. Like many if not most channels, the ample presence of sand and low vegetation cover create conditions ripe for frequent aeolian transport. Additionally, since any event that brings water and sand down the channels also tends to bring PM_{10} producing material, the PM_{10} emission potential of these channels can be quite high. It is therefore not surprising that PM_{10} emissions have been observed from the wash/fan complexes, especially after recent flooding caused by heavy rains from atmospheric rivers and tropical storm events (Figure 4-17; see also discussion of Shell Cut monitoring site in Chapter 3). Although flooding at Centennial Wash in 2022 and 2023 (Figure 4-18) generated more dispersed flood deposits than at Coso Wash, analysis of pollution rose data and other data in Chapter 3 suggest that Coso Wash was the dominant cause of elevated PM_{10} at the Shell Cut monitoring on the southern side of the lake. Therefore, the committee more closely examined the Coso Wash area.

Such flood events have little to do with anthropogenic activities. However, to the extent that climate change affects the frequency and intensity of these large, rare events, the emissions that they cause may be expected to increase or decrease. The panel's analysis of aerial and satellite imagery of the Coso Wash (Figure 4-17), shows little in the way of anthropogenic impacts that might lead to increased PM_{10} emissions from the fan. No evidence was found, for example, of transport of sand from the exposed lakebed onto the fan, where it may have either changed the morphology of the fan/channel or where it may have provided excess saltators in the presence of PM_{10} material. Although some highway berms or ditches were constructed in the past to reduce flood impacts to Highway 190, analysis of aerial imagery in Figure 4-17 does not suggest that any structures have altered flood flows in the Coso Wash in ways that would be likely to increase PM_{10} emissions.

Dirty Socks Area

No evidence was found in the Dirty Socks area (Figure 4-19) of impacts that may have increased PM_{10} emissions related to the drainage of Owens Lake. However, the panel did note that there is an area of flood deposits sitting behind a beach ridge just south of the Dirty Socks monitor. In 1944, this appeared as a relatively small area with Highway 190 cutting across one side. By 1977, this highway had been moved to the higher beach ridge to the north. From 1977 onward, the area of flood deposits behind the beach ridge has continued to grow, representing successive flood events that deposited material behind the beach. These deposits are likely possible sources for exceedances observed at the Dirty Socks station, which is immediately to the north. Although these images are not definitive, there is a possibility that moving the highway increased the potential of this small depression to store impounded water and sediment. In doing so, it may have inadvertently increased the potential of this area to contribute to PM_{10} emissions. Detailed high resolution imagery and coring for geochronology and geochemistry could further constrain the impact the highway had on the flood deposit. In particular, coring for ages of sediment deposition using OSL could establish the cycles of flooding, the amount of sediment tied to the flood events, and fine-grained storage potential of the Dirty Socks location. This sampling would be most effective if it occurred in conjunction with a well-defined grain size and geochemical analyses of each stratigraphic unit.

Olancha Dunes

The Olancha Dunes overlie late Pleistocene and Holocene alluvial fan and lake deposits associated with late Holocene hydroclimate variations at Owens Lake. The Olancha Dunes are the primary dune area in the southern sector of the Owens Lake basin (Lancaster et al. 2015) and consist of approximately 4.39 km^2 of sparsely vegetated low dunes with isolated ridges that extend nearly 5 km southwest from the historical shoreline of Owens Lake. The Olancha Dunes are composed of poorly to moderately sorted fine to medium-sized sand (Figure 4-4B; Lancaster

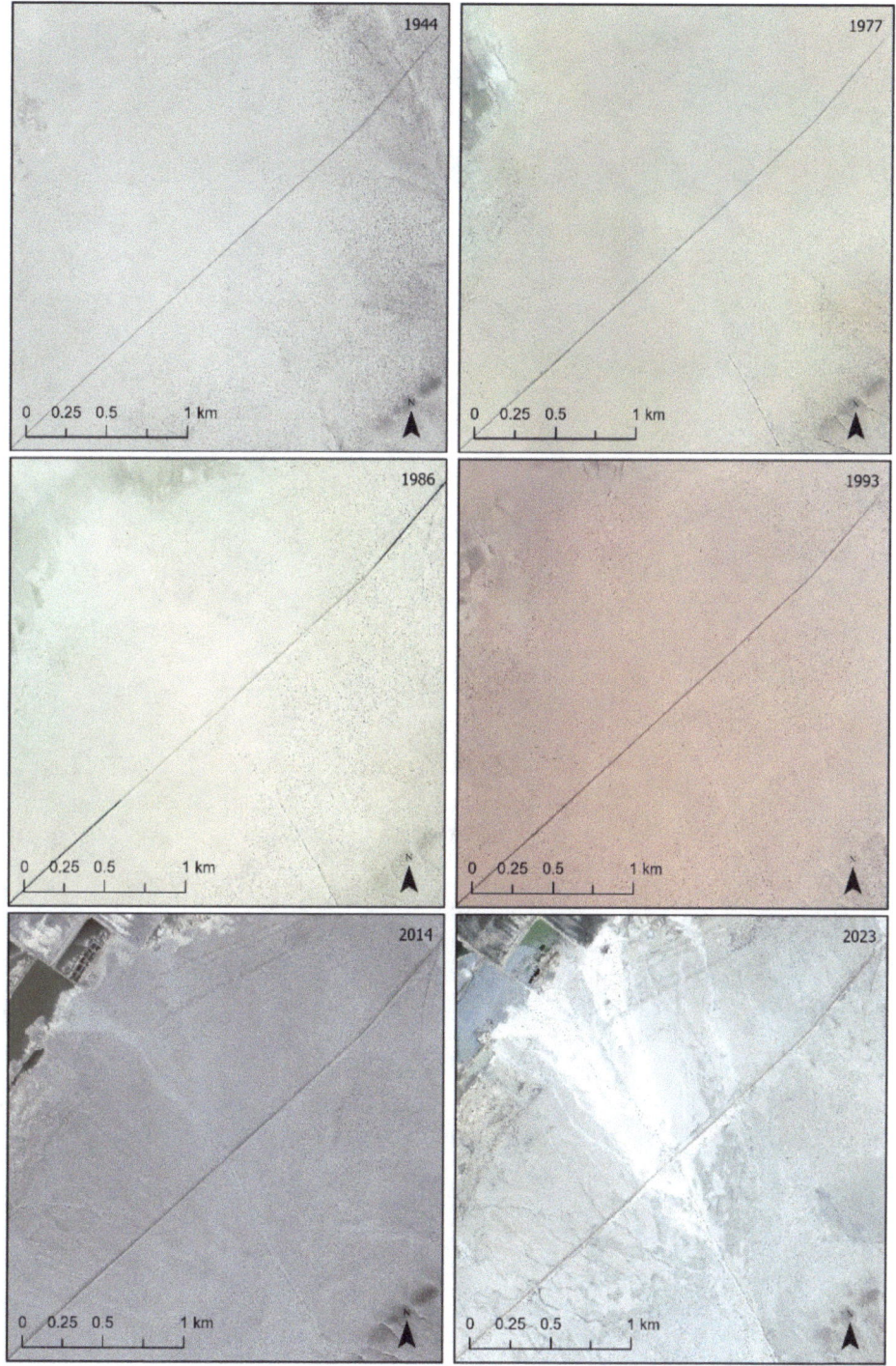

FIGURE 4-17 Images of the Coso Wash and fan area.
NOTES: Highway 190 is visible in 1944, 1977, 1986, 1993, 2014, and 2023. The pathways of floodwaters are highly visible in 2023 after the remnants of Hurricane Hilary in 2023. The panel did not find any evidence of anthropogenic impacts that might lead to increased PM_{10} emissions from the fan.
SOURCE: Aerial images provided by LADWP.

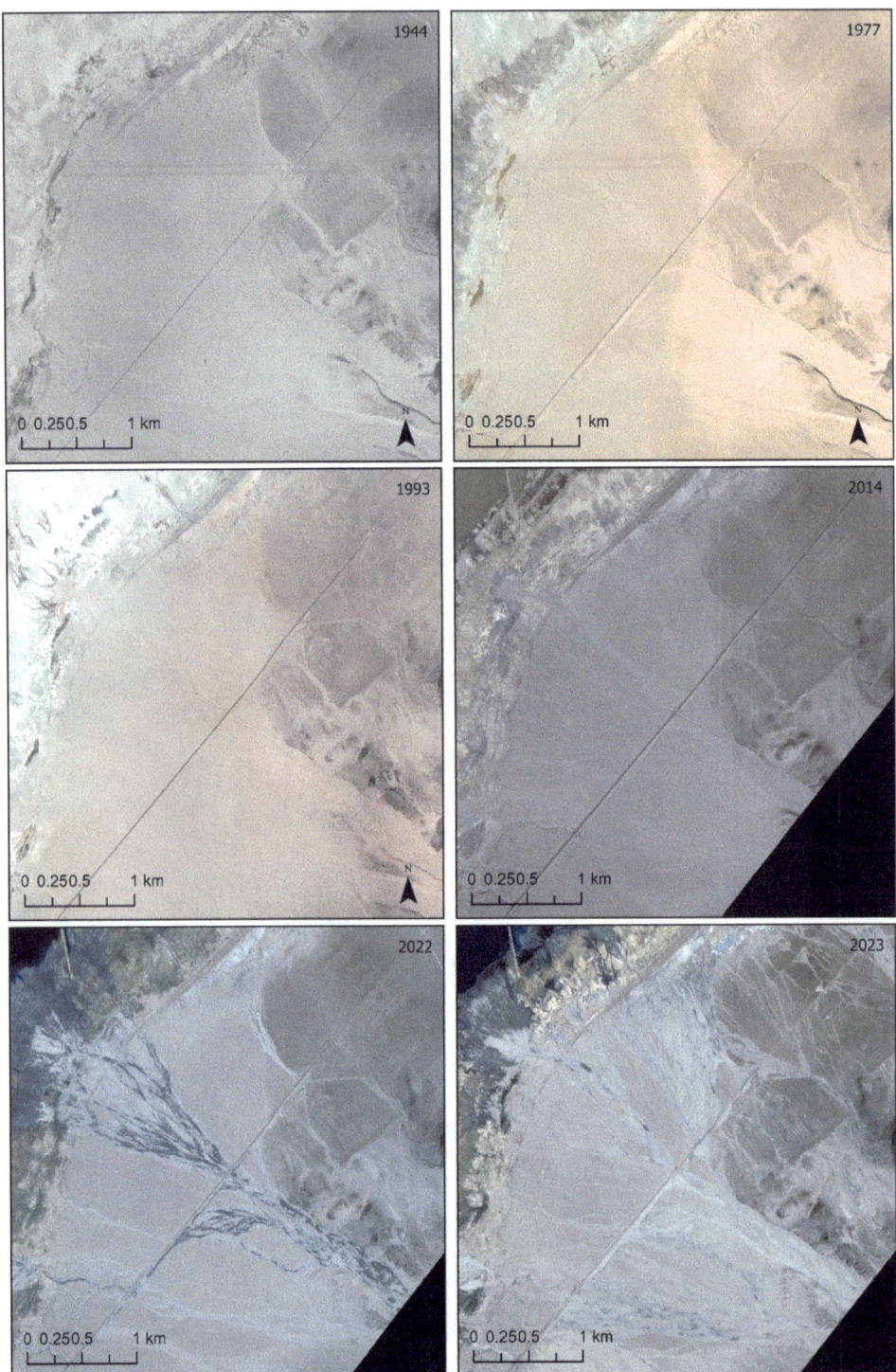

FIGURE 4-18 Images of the Centennial Wash and fan area.
NOTES: Highway 190 is visible in 1944, 1977, 1993, 2014, 2022, and 2023. The pathways of floodwaters are highly visible in 2022 after Tropical Storm Kay and in 2023 after the remnants of Hurricane Hilary and show how highway infrastructure alters flow paths.
SOURCE: Aerial images provided by LADWP.

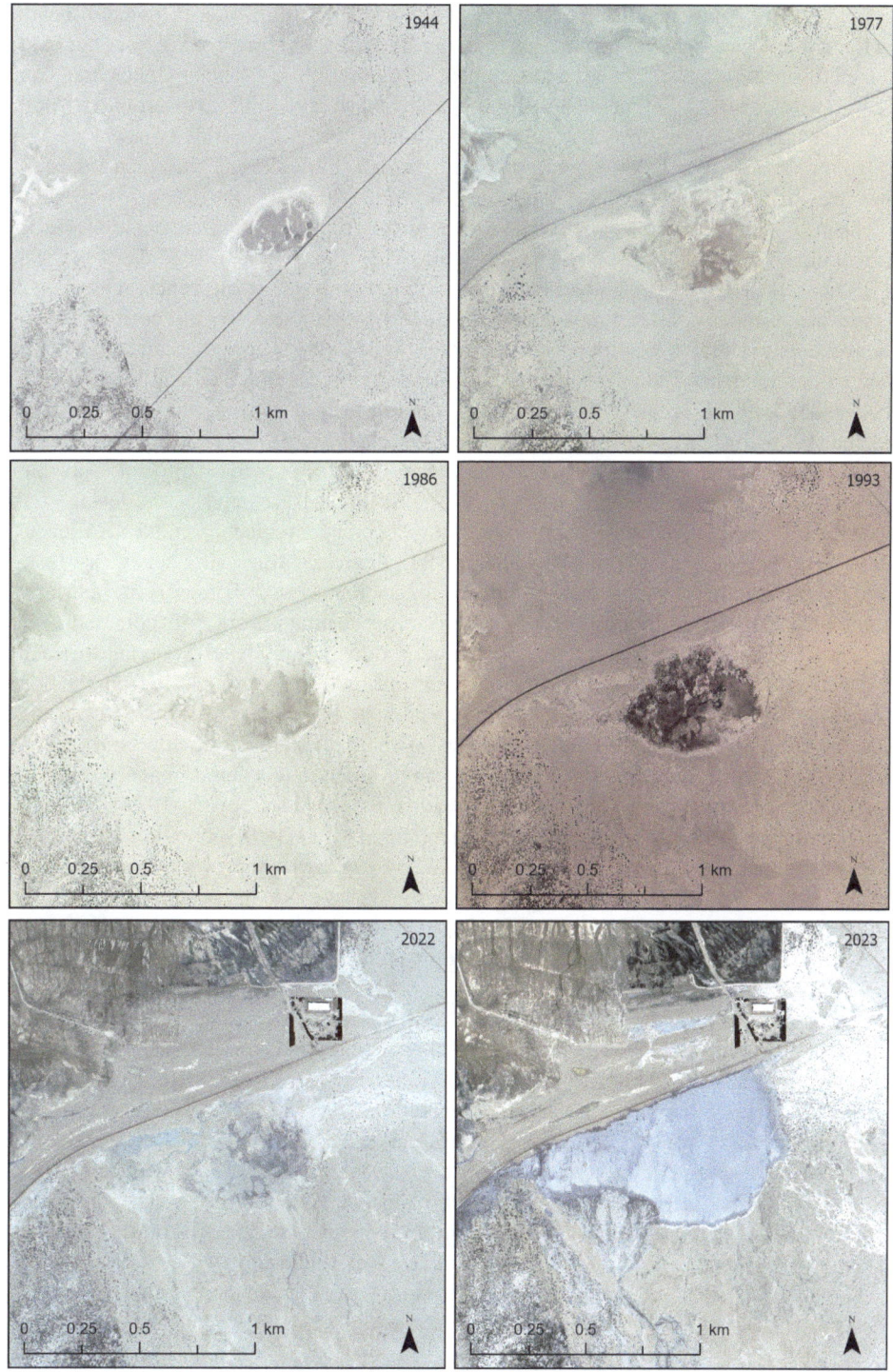

FIGURE 4-19 Aerial and satellite images of the area around Dirty Socks showing a small area of flood deposits behind beach ridges.
NOTES: Between 1944 and 1977, Highway 190, which had originally gone through these flood deposits, was rerouted to sit on top of the beach ridges. This appears to have been related to an increase in the size of the flood deposits backing the beach/road.
SOURCE: Aerial images provided by LADWP.

et al. 2015). Sand from the southern portion of the Owens Lake basin is mineralogically mature, with the Olancha Dunes having the highest quartz content compared to other dune areas (Figure 4-4A; Lancaster et al. 2015). As at the Keeler Dunes (discussed above), the high enrichment of quartz in the Olancha Dunes area can be attributed to either abrasion of feldspars during aeolian transport or weathering of feldspars during transport or phases of dune stabilization (Muhs 2004) followed by immersion and transport in lacustrine environments during historic water-level fluctuations in Owens Lake. Sand in the southern part of the Owens Valley basin, especially near the Olancha Dunes, was likely sourced from the South Sand Sheet (Figure 4-10) during prehistoric low stands of the Owens Lake. The South Sand Sheet is hypothesized to be sourced from a system of ephemeral washes draining the Coso Range, including Vermillion Canyon (Lancaster et al. 2015). Bacon et al. (2020) dated two paleo-shorelines at elevations of 3,638 ft (1,109 m) and 3,635 ft (1,108 m) and obtained luminescence ages of 11.5–45.6 ka. To date, we are unaware of other constraining ages. Using augers to complete an OSL coring campaign across the sand dunes/sand sheets could also provide a comprehensive set of young ages that are reflective of recent dune activation. While quartz would be the primary target of these studies (as the dominant mineral and best suited to produce higher-resolution OSL ages), feldspar may also provide helpful OSL ages and distinctive luminescence characteristics that allow for increased confidence in the OSL chronology.

The resultant aeolian drift potential derived from recent decades of wind data near Olancha (Lancaster et al. 2015) indicates a strong bimodal sand transport regime at the southern end of the Owens Lake basin with opposing northerly (29 percent frequency) and southerly (51 percent frequency) modes. In recent decades, net resultant drift at Olancha is toward the north/northwest (345 degrees), although a secondary reversed transport mode toward the south/southwest also occurs. Similarly, the aeolian sand drift regime at the nearby Dirty Socks site indicates bimodal transport with a northerly (2 degrees) net resultant sand drift potential vector, but with a more balanced frequency of opposing northerly and southerly transport modes (42 percent for each). As a result, the present connection between sediment supply from the Owens Lake bed, particularly from the South Sand Sheet area (Lancaster et al. 2015) to lake-marginal dunes at Olancha, is not as clear as it is on the northeastern side of the lake at Keeler. Although the resultant sand drift potential vector and crescentic dune slip faces in the Olancha Dunes indicate net northward sand transport, it is also apparent in historical aerial imagery (Figure 4-20) that the dune field has expanded toward the south, particularly since the 1970s. Between 1977 and 1986, unvegetated areas of the southeastern portion of the dune field expanded approximately 0.3 miles south toward Fall Road and a private residence evident in the imagery (see area enclosed by yellow line in 1986 image in Figure 4-20). It is possible that this trend partly reflects the secondary south/southwest mode in the sand transport regime resulting from increased sediment supply or availability. This sediment supply or availability could have resulted from the drained lakebed or other anthropogenic activities at the Olancha Dunes, but this cannot be distinguished with existing monitoring data.

The 1970s and 1980s correspond to an era of increasing off-highway vehicle (OHV) activity within dune fields in California both generally and specifically in large recreational preserves, such as the Oceano Dunes State Vehicular Recreation Area and the Imperial Sand Dunes Recreation Area. Research at these sites shows clear associations between OHV activity, decreasing vegetation cover, and increased dust emissions (Cheung et al. 2021; Gillies et al. 2022; Groom et al. 2007; Walker et al. 2023). In the Owens Valley, the Bureau of Land Management (BLM) manages an OHV recreation and primitive camping area of approximately 400 acres (1.6 km^2) or 36 percent of the total area of the Olancha Dunes, which has been operational for decades. Although there is little to no research on the trends in vegetation cover within the Olancha Dunes or on the associated impacts of decades of OHV activity on dune development and dynamics, it is probable that the extent and duration of OHV activity at the Olancha Dunes could explain recent declines in vegetation cover and dune expansion, particularly in the southern area and other areas of active camping and riding.

NORTH OF THE LAKE

As shown in Chapter 3, there have been a few exceedances observed at Lone Pine or other northerly monitors like North Beach and Lizard Tail between 2017 and 2024. The District's exceedance database suggests that many of the off-lake exceedances at these monitors are from anthropogenic disturbances (e.g., landfill) or from regional sources

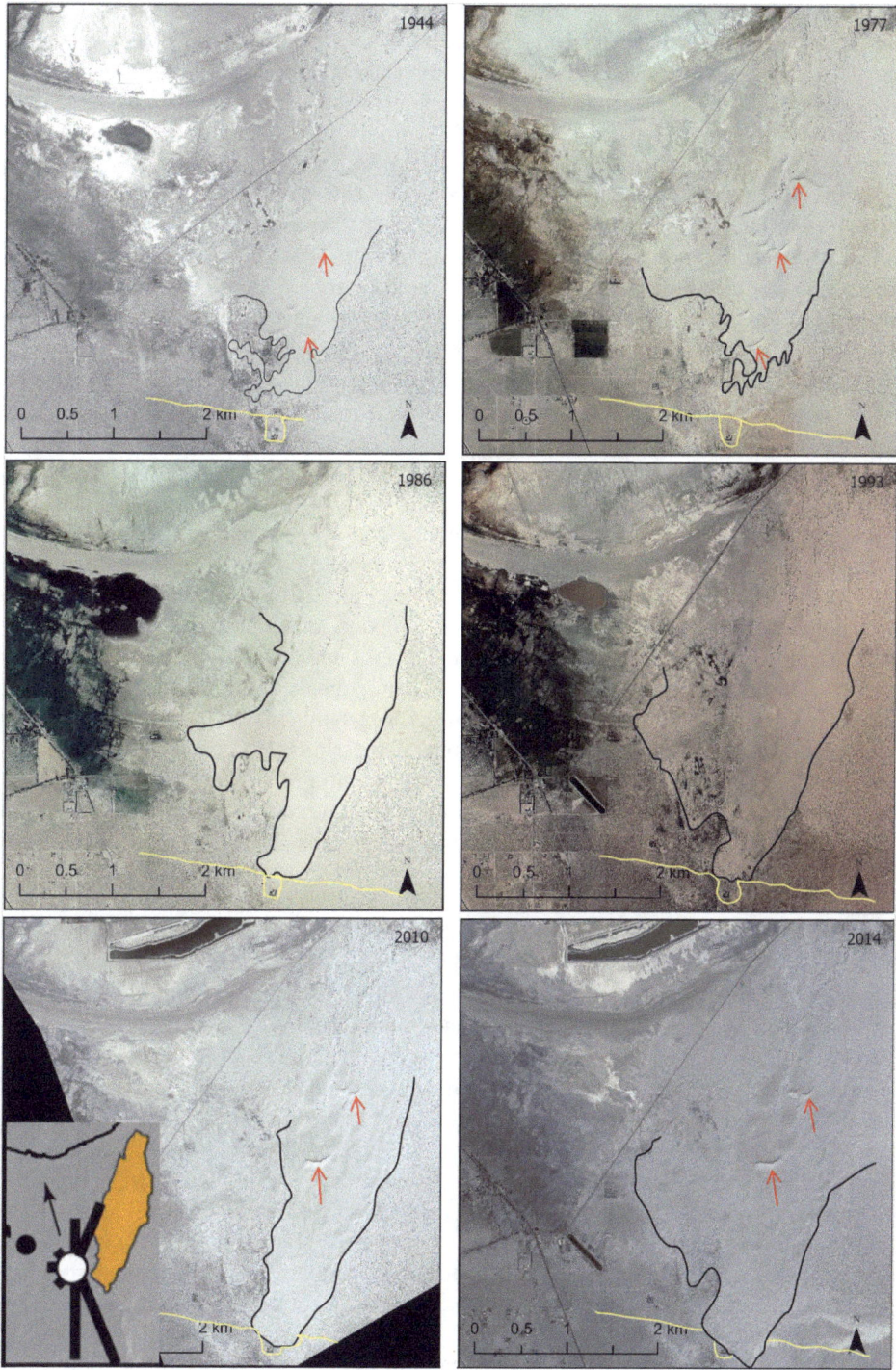

FIGURE 4-20 Historical aerial photography of the Olancha Dunes region south of Highway 190.
NOTES: Arrows indicate formative wind directions for nearby crescentic dune ridges, all of which appear to have been formed by southerly winds. Black outline is estimate of active, largely unvegetated portions of the dune field. Yellow lines indicate an established road and property at the southern end of the dune field.
SOURCE: Aerial images provided by LADWP. Wind rose from Lancaster et al. (2015).

outside the OVPA. However, some evidence shows that certain regional events were augmented by sources north of the lake. These include a regional exceedance on 05/11/2018 that was "augmented by local sources north of North Beach between the monitor and the Lone Pine landfill," a regional exceedance on 10/27/2019 that was "augmented by local sources north of North Beach monitor," and a regional exceedance on 10/11/2021 that was affected by "additional sources as the front traveled down the valley." Additionally, as demonstrated in Chapter 3, sources near the Fort Independence monitoring site may be primary causes of a few high PM_{10} concentrations observed there.[4] However, as discussed in Chapter 3, it is unclear if these sources are derived from active disturbances of the surface (e.g., construction, highway traffic), or if they are from other off-lake sources of dust that are the focus of this report. Given that the panel was tasked to consider "expected future changes in off-lake dust sources over time," potential future sources of dust emission to the north of the lake are explored here, even though it is not clear that these sources are currently resulting in exceedances. These potential sources include small depressions found inset in the land surface, vegetated areas found throughout the valley floor including agricultural fields (active and abandoned), as well as areas of former alkali meadow between Lone Pine and the north edge of the OVPA.

Microplayas

A considerable number of small areas, mapped as "microplayas" by the District, exist in the area around North Beach within 3 km of the historical shoreline (Figure 4-21). These mapped areas likely do not all share the same origin; they may be depositional, erosional, or a combination of the two. For example, some could be paleo-lake deposits that have been exhumed by wind (Stone et al. 2000). Alternatively, these landforms could be wind erosion hollows that subsequently filled with heavy-texture material from local overland flow (Stone et al. 2000). Other examples of these landforms sit at the toe of fans and are likely purely depositional features. At present, there is no evidence to show that microplayas are anthropogenic in origin, but they could be classified as such if human activity led to the wind erosion (e.g., through vegetation disturbance or groundwater disturbance that impacted vegetation) that created these local erosional depressions. Nonetheless, amid otherwise relatively coarse soils, these landforms do have more potential to contribute to PM_{10} exceedances compared to other areas of the uplands, especially because they appear to have lower vegetation cover compared to the surrounding areas. Nonetheless, there are currently no exceedances obviously associated with these surfaces, and it is unlikely that they will be major source of exceedances in the future.

Vegetated Areas North of the Lake, Including Areas of Former Alkali Meadows

As described in Chapter 2, large alkali meadows historically existed throughout the Owens Valley (Benson et al. 2002), but much of this groundwater-dependent vegetation was lost, and at present, the dominant vegetation type in the Owens Valley is characterized as arid or semiarid scrub (LADWP and County of Inyo 1990a). When loss of groundwater-dependent vegetation like this occurs, meadow vegetation is frequently replaced by desert shrublands dominated by xeric shrubs such as saltbush (*Atriplex* spp.), sagebrush (*Artemesia tridentata*), and blackbrush (*Coleogyne ramisissima*). The result is a conversion of a meadow with greater than 30 percent cover (often considerably more) to scrublands with 10–20 percent cover (Figure 2-5; Manning 1997). Vest et al. (2013) showed nearly an order of magnitude increase in aeolian transport between alkali meadow and scrub sites, and the fluvial and lacustrine sediments across the OVPA valley floor (Bacon et al. 2006) provide ample saltation and suspension-sized material. Thus, the panel investigated the potential for historical alkali meadows to produce PM_{10}, even though based on the analysis of data in Chapter 3, it is not certain that this is currently resulting in exceedances at existing monitoring sites.

As described in Chapter 2, historical anthropogenic activities such as the increase in groundwater pumping, and surface water in-valley use and export to Los Angeles after 1970 impacted vegetation along the OVPA valley floor. Between 1986 and 1992 there was a considerable but variable decrease in groundwater levels in the Owens Valley

---

[4] The U.S. Environmental Protection Agency (EPA) does not use this monitor to assess attainment with the PM_{10} National Ambient Air Quality Standards (NAAQS).

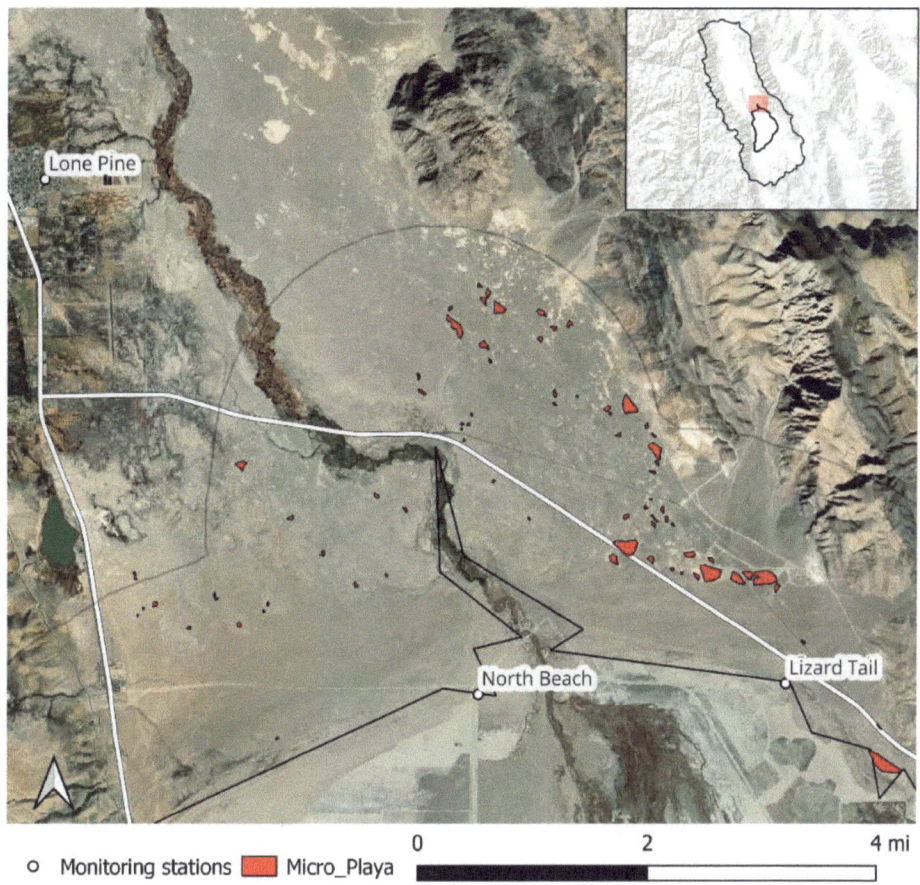

FIGURE 4-21 Microplayas that have been mapped by the District north of Owens Lake.
NOTES: The dark black line is the 3,600-ft regulatory shoreline and the light grey line delineates an approximate 3 km distance from the shoreline.
SOURCE: Data from Nik Barbieri, GBUAPCD, personal communication, July 2024.

as a result of this use (Vest et al. 2013), which was associated with significant vegetation change at those locations that later required revegetation projects to reduce dust emissions (LADWP 2023). For example, LADWP (LADWP 2023 pp. 3-14 to 3-15) states, "Fluctuations in water tables due to groundwater pumping have caused approximately 655 acres of groundwater-dependent vegetation to die off [at Hines Spring South]. …The goal will be to restore as full a native vegetation cover as is feasible, but at a minimum, vegetation cover sufficient to avoid blowing dust will be achieved in that area." Further evidence of historic emissions induced by groundwater withdrawals was found in the highly wind-eroded Rindge soils that were visited by the panel during the May 29–30, 2024, meeting south of Independence. Evidence of decimeters of historic erosion by wind was evident in a location that was at the site of a former groundwater seep, as evidenced by the high organic matter content and clear histic epipedon. Thus, the panel finds that areas north of the lake within the OVPA, including historical alkali meadows, have produced dust in the past and thus have the potential to produce dust in the future.

Irrigated agriculture removes native vegetation and changes the soil structure, which can lead to dust emission, especially if these agricultural lands become abandoned (Birmili et al. 2008; Field et al. 2010). As described in Chapter 2, there was a reduction in the amount of irrigated acreage of Los Angeles–owned land from the mid-1960s to 1970 (LADWP and County of Inyo 1990b). Abandoned agricultural land does not have the same cover of native perennial vegetation as similar areas that were not plowed, and thus they generally produce greater amounts of

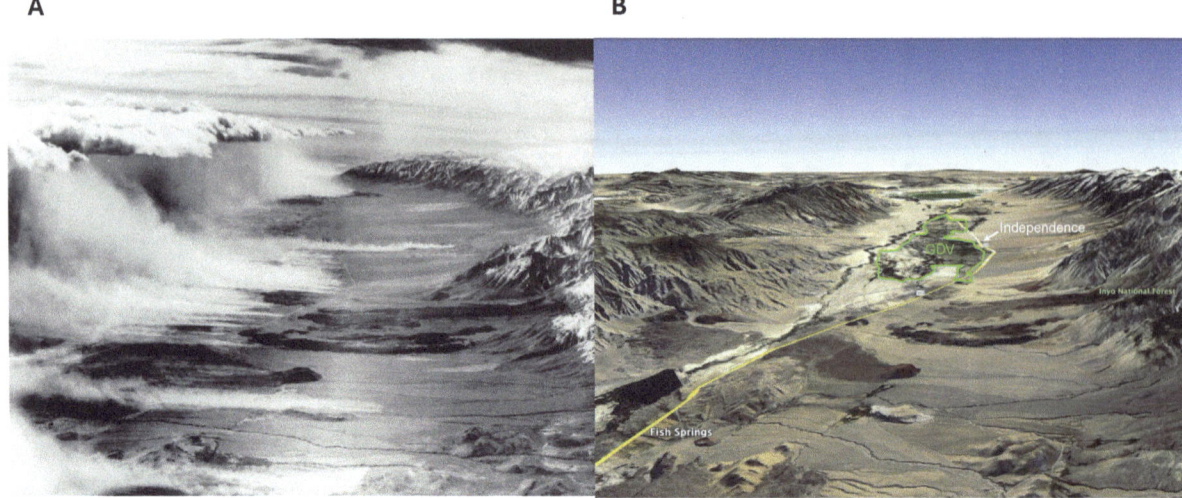

FIGURE 4-22 A) Oblique aerial photo of blowing dust over the Owens Valley on March 5, 1950, looking south. Notably, dust emissions are not seen in many areas of remnant groundwater-dependent vegetation (GDV) west of the Owens River in the vicinity of Independence; B) Similar view from Google Earth, with GDV delineated.
NOTES: The location of Highway 395 was moved after 1950. Owens Lake can be seen in the south.
SOURCES: A) Photograph by Robert Symons; B) Google Earth.

dust (Zucca et al. 2022). A photograph from 1950 (Figure 4-22A)[5] shows extensive dust emission from the area north of Owens Lake in the vicinity of Independence, likely from disturbed land or active, fallowed, or abandoned agricultural fields.

A groundwater management plan was implemented in the 1990s (see Chapter 2) along with revegetation projects on abandoned agricultural land and land that was impacted by the loss of groundwater-dependent vegetation. LADWP (2023) notes that many of the revegetation projects are "implemented and ongoing," whereas others are "complete," and still others are "fully implemented but not meeting goals." However, the panel did not attempt to determine if the completed projects have sustained their target vegetation cover nor whether these levels are sufficient to suppress dust emission.

As described in Chapter 2, the panel expects considerable future change in climate in the region of the Owens Valley. Warming will drive a large loss of snowpack, a dramatic shift in runoff time to earlier in the wet season, and an increasing "flashiness" of runoff (Alex Hall, personal communication, July 2024; Harpold et al. 2015). Measured and projected increases in temperature (Abatzoglou et al. 2021; Williams et al. 2019) have and will continue to increase evaporative demand, affecting all types of vegetation. Increasing evapotranspiration will drive reductions in surface runoff, even in scenarios with increasing precipitation (Owens Valley Groundwater Authority 2021). Future changes in precipitation are predicted to bring longer-term drought conditions with intermittent extreme wet events (Swain et al. 2016; 2018). Extended drought could impact areas of natural non-groundwater-dependent vegetation, areas of former groundwater-dependent vegetation, or former irrigated agricultural fields, because rainfed vegetation remains susceptible to drought. Projecting into the future, a combination of climate changes (e.g., decreased runoff, increased evapotranspiration, increased drought) could result in increased dust emission from vegetated areas north of the lake, especially if there are changes in groundwater management agreements or land management policies. Revegetation projects that have not yet met their revegetation goals may be especially susceptible to future changes in climate because such areas may have soil/climate conditions that do not favor rapid revegetation; in such places, a climate less amenable to vegetation growth is likely to slow progress even further.

---

[5] See https://www.soaringmuseum.org/hof_more.php?id=117.

CONCLUSIONS AND RECOMMENDATIONS

The panel considered multiple lines of evidence to infer the origins and evolution of major PM_{10} sources in the OVPA or those that might become important sources in the future. One important process the panel considered was "winnowing." This process suggested that PM_{10} material from the dry lakebed that may have been deposited onto off-lake landforms would be expected to decrease over time due to its resuspension and progressive removal by aeolian processes. This hypothesis was based on a correlation between on- and off-lake exceedances at the Dirty Socks monitor between 1999 and 2012, which has not been borne out with more recent data for estimated emissions trends and exceedances. Most current PM_{10} emissions from off-lake areas are likely not a result of resuspension of PM_{10} material that was deposited on off-lake landforms from the lakebed. Instead, the presence and common replenishment of highly emissive flood deposits provides ample fine particulates that can be emitted as PM_{10} as long as the horizontal flux of saltation-sized particles is sufficient to emit dust from the surface.

Conclusion 4-1: Winnowing is expected to play a minimal role in reduction in future off-lake PM_{10} exceedances.

Northeast Side of the Lake

The northeastern side of Owens Lake is host to several landforms including the Keeler Dunes and the Slate Canyon/Keeler Alluvial Fan Complex that have overgone major changes over the last century and are substantial contributors to exceedances. During the 20th century, the Keeler Dunes transitioned from a largely vegetated dune system that was stabilized by greasewood (*Sarcobatus vermiculatus*) to an active dune field. The emergence of Keeler Dunes as an active dune field resulted in abundant saltation that can drive PM_{10} emissions from flood deposits that are continually replenished from the alluvial fan. Therefore, ongoing PM_{10} exceedances from this area are a direct result of the destabilization of the Keeler Dunes. The panel finds that the net transport direction, available imagery, and the evidence for groundwater-dependent vegetation currently present in the dunes supports the conclusion that increased sand transport following the diversion of water from the Owens Lake destabilized the Keeler Dunes. Changes to surface hydrology resulting from construction of berms above Highway 136 appear to have had an impact on upland (non-groundwater-dependent) vegetation but are unlikely to have led to the destabilization of groundwater-dependent vegetation in the Keeler Dunes.

Conclusion 4-2: The reactivation of the Keeler Dunes was related to the additional upwind sand supply available from the Owens River delta following drainage of Owens Lake.

Conclusion 4-3: Due to continuing aeolian activity of the Keeler Dunes and replenishment of flood deposits within the dunes, the system will continue to contribute material to PM_{10} emissions. Stabilization of the dunes would likely reduce PM_{10} emissions.

Several constructed berms northeast of Highway 136 were intentionally designed to alter surface hydrology, directing overland flow to specific points of discharge points along the highways. The panel did not analyze each berm-related flood deposit, but instead considered the berms on the Keeler/Slate Canyon Fan as potentially representative of similar features around Owens Lake. These berms have had appreciable, localized impacts on the distribution of flood deposits in the Keeler Dunes region, especially following impacts from the remnants of Hurricane Kay in September of 2022. Further investigation would be needed to determine the impacts that these berms have on the potential for off-lake PM_{10} emissions from flood deposits, if any.

Conclusion 4-4: The construction of berms above Keeler Dunes and elsewhere modified sediment transport, but it is uncertain if this modification of sediment transport increased PM_{10} emissions from flood deposits relative to that which would have occurred without berms.

Recommendation 4-1: The District should work with the California Department of Transportation and other Owens Valley Planning Area landowners to determine the impact of berms on flood deposits and associated PM_{10} emissions.

Southern Side of the Lake

The southern side of Owens Lake is host to the Olancha Dunes and multiple alluvial systems that are important sources of PM_{10} exceedances. The scientific literature on the origin and evolution of the Olancha Dunes is quite sparse, but what available evidence does exist indicates that the dunes formed prior to the diversion of water from Owens Lake. The panel's analysis shows that the dunes experienced a slight southward extension (approximately 0.3 miles) from 1944 to current day. This southward extension could be the result of increased sediment supply following the diversion of water from Owens Lake or from other natural or anthropogenic activities. Olancha Dunes is also the location of an OHV and dispersed camping recreational area that makes up approximately 36 percent of the total dune area. There is little to no research on the impacts of decades of OHV activity on PM_{10} emissions at Olancha Dunes, but research from other sites like the Oceano Dunes State Vehicular Recreation Area and the Imperial Sand Dunes Recreation Area shows clear associations between OHV activity, decreasing vegetation cover, and increased dust emissions. Additional study using aerial photography, Portable In-Situ Wind Erosion Lab dust emission potential measurements, and the Bureau of Land Management's records of impacts from recreational activity could provide information on the contribution of recreational activity to PM_{10} emissions.

Conclusion 4-5: The Olancha Dunes has extended southward slightly since the 1940s, but there is not sufficient evidence to indicate that this southward extension was influenced by drainage of the lake or other anthropogenic activities, such as OHV recreation and dispersed camping.

Recommendation 4-2: The District should work with the Bureau of Land Management to determine the impacts of recreational activity on plant communities and PM_{10} emissions within the Olancha Dunes and remediate as needed.

The southern side of Owens Lake also hosts multiple alluvial channel/wash systems that deliver and rework sediments from the neighboring Coso and Sierra Nevada ranges. These alluvial channel/wash systems supply sand and PM_{10} material, and they can only support low-density vegetation cover, which create conditions ripe for high PM_{10} emission. While the replenishment of these alluvial systems is a natural process that has been occurring for millennia, anthropogenic alteration of the flowpaths through the constructed infrastructure may change the amount and distribution of impounded water and sediment and therefore change its potential to contribute to PM_{10} emissions. Climate change is projected to make extreme precipitation events more frequent and intense, which would more frequently replenish fine sediments in flood deposits that contribute to PM_{10} emissions.

Conclusion 4-6: Aerial and satellite images suggest that the impounded flood deposits south of the lake near the Dirty Socks PM_{10} monitor may have been affected by the rerouting of Highway 190. Highway 190 infrastructure clearly impacts flood flows in other areas along the south of the lake, although it is unclear to what extent, if any, this infrastructure impacts overall PM_{10} emissions and measured exceedances.

Recommendation 4-3: The District should work with the California Department of Transportation to determine the impact of Highway 190 and related berms on flood deposits and associated PM_{10} emissions, with initial emphasis on the impounded flood deposits near the Dirty Socks PM_{10} monitor.

POTENTIAL SOURCES NORTH OF THE LAKE

Current data indicate a stable shallow groundwater table in the area around Owens Lake. However, there is substantial evidence that areas to the north of the lake have seen decreases in vegetation cover, which may have contributed to historical dust emission in the OVPA. A groundwater management plan and a number of revegetation projects were implemented in the 1990s to reduce blowing dust in affected areas. If this land is not managed carefully, dust emission from the area to the north of the lake could increase, especially under changing climate conditions.

Conclusion 4-7: Drought coupled with constant or increasing water extraction in the Owens Valley could result in prolonged lowering of the groundwater table. If groundwater drops to levels that severely impact the health of existing groundwater-dependent vegetation, the potential for PM_{10} emissions north of the lake would increase.

Conclusion 4-8: Continued monitoring and regular updates on the advancement of revegetation projects on former groundwater-dependent meadows and abandoned agricultural fields will inform potential measures that may be necessary to reduce PM_{10} emission in the face of future climate pressures.

FURTHER RESEARCH TO ESTABLISH THE ORIGIN AND EVOLUTION OF OVPA SOURCES

More chronological research may reduce uncertainties surrounding the origin and evolution of dune fields and flood deposits. Collecting sediment samples across dunes and flood deposits by coring or auguring may be the best way to collect a comprehensive set of data. These methods may illuminate relatively recent processes that occurred after the diversion of water from Owens Lake, construction of berms, and the rerouting of Highway 190.

Conclusion 4-9: A coring and optically stimulated luminescence campaign targeting recent mobilization events (including those younger than 100 years) across Olancha Dunes, Keeler Dunes, and the flood deposits near the Dirty Socks monitor will reduce uncertainty on the origin and evolution of these deposits.

5

Utilization of the U.S. EPA Exceptional Events Rule in the OVPA

This chapter discusses the applicability of the U.S. Environmental Protection Agency's (EPA's) Exceptional Events Rule (EER; 40 C.F.R. § 50.14) for excluding air quality monitoring data when determining the Owens Valley Planning Area's (OVPA's) compliance with the National Ambient Air Quality Standards (NAAQS; Box 1-1). Based on the statement of task and information provided by the sponsors, this chapter focuses only on PM_{10} exceedances that were identified by the Great Basin Unified Air Pollution Control District (the District) as originating from local off-lake sources (i.e., those OVPA sources that are above the 3,600-ft-elevation regulatory shoreline) and the necessary criteria to demonstrate exceptional events due to high wind dust events. However, the panel recognizes that on any individual monitored exceedance day, dust from both on-lake and off-lake sources may contribute to PM_{10} exceedances at monitoring sites in the OVPA to varying degrees. This chapter does not consider exceptional event demonstrations for wildfires or regional events, nor does it include findings from the agencies able to prepare and/or concur with exceptional events demonstration(s).[1] This chapter also does not provide approvable analysis results under the EER for the OVPA; only the air quality management agencies (hereafter air agencies) can make EER-applicable demonstrations as authorized under the federal Clean Air Act.

THE EXCEPTIONAL EVENTS RULE

An exceptional event is "an event and its resulting emissions that is not reasonably controllable or preventable, and is caused by human activity that is unlikely to recur at a particular location or a natural event" (40 C.F.R. § Part 50.1[j]). Exceptional events are considered to be "events for which the normal planning and regulatory process established by the Clean Air Act (CAA) is not appropriate" (Treatment of Data Influenced by Exceptional Events, 72 FR 13560, 2007). When an air agency demonstrates (and EPA concurs) that a PM_{10} exceedance is an exceptional event, those data are excluded from specific decisions determined to be of "regulatory significance" to be made under EPA CAA authority related to NAAQS attainment and related planning requirements. These regulatory significant decisions include NAAQS area designations and redesignation decisions, classifications, determinations regarding whether a nonattainment area has reached attainment, certain findings of State Implementation Plan (SIP) inadequacy, and other actions on a case-by-case basis (40 C.F.R. § 50.14[a][1]). Due to the 3-year design value period, it often takes more than 3 years to determine if an EER demonstration would have the effect of changing

---

[1] For the OVPA, these agencies are the U.S. EPA, the California Air Resources Board (CARB), and the Great Basin Unified Air Pollution Control District (GBUAPCD).

the rolling 3-year average of exceedances used to determine attainment at an individual monitor or monitors and whether that effort could support a change in nonattainment status or affect related planning requirements.

Although not necessarily intuitive from the name, the determination of exceptional events is unrelated to event frequency (or infrequency). Instead, in the case of high wind events, certain exceedance event data can be excepted (i.e., excluded) by demonstrating the following criteria:

- the high-wind event clearly caused or led to the monitored exceedance,
- the exceedance event "was both not reasonably controllable and not reasonably preventable," and
- the exceedance "was caused by human activity that is unlikely to recur at a particular location or was a natural event," among other requirements, including public notice (EPA 2019b, p. 2).

EPA (2019b, p. 4) defines the high wind threshold as "the minimum wind speed capable of causing particulate matter emissions from natural undisturbed lands in the area affected by a high wind dust event." As specified in the Exceptional Events Rule, EPA accepts a threshold of a sustained wind[2] of 25 miles per hour (mph) for certain areas of the western U.S..." including California, "as long as the value is not contradicted by evidence in the record at the time the demonstration is submitted." For California, the high wind threshold of a sustained wind of 25 mph is based on findings from James et al. (1999) and Wacaser et al. (2006).[3] However, the EER specifies protocols for evaluating any given exceedance event based on three tiers: Tier 1) for high-energy dust events with ≥40 mph sustained wind speed and ≤0.5 mile visibility, Tier 2) for events with ≥25 mph sustained winds, and Tier 3) for all other high wind dust events with <25 mph sustained wind speeds. For events with lower wind speeds, increasing evidence is required to justify an exceptional event (EPA 2019b).

The EER defines a natural event as "an event and its resulting emissions, which may recur at the same location, in which human activity plays little or no direct casual role" (40 C.F.R. § 50.1[k]). EPA (2019b, p. 16) further clarifies that "an event involving windblown dust solely from natural undisturbed landscapes is considered a natural event and therefore not reasonably controllable." Thus, according to EPA guidance, any PM_{10} exceedance event caused by a source judged to be natural meets the definition as "both not reasonably controllable and not reasonably preventable." EPA generally considers natural disasters (e.g., hurricanes, wildland fires) to be natural events (Treatment of Data Influenced by Exceptional Events, 72 FR 13560, 2007), even though some of them may have been exacerbated by human activities and human-induced climate change. Additionally, the rule states: "The Administrator will consider high wind dust events to be natural events in cases where windblown dust is entirely from natural undisturbed lands in the area or where all anthropogenic sources are reasonably controlled...." (40 C.F.R. § 50.14[b][5]). Thus, events that include emissions from mixed sources (i.e., both natural and anthropogenic sources) can be considered natural events only if reasonable controls have been applied to the contributing anthropogenic sources (Treatment of Data Influenced by Exceptional Events, 72 FR 13560, 2007).

As stated in 40 C.F.R. § 50.14(b)(8), an event from an anthropogenic source is considered not reasonably controllable if "reasonable measures to control the impact of the event on air quality were applied at the time of the event." According to 40 C.F.R. § 50.14(b)(5)(iv), air agencies are not required to submit justification for a high wind dust event not being reasonably preventable. Documentation of Tier 1 high-energy dust storms (≥40 mph sustained wind speed) is generally considered sufficient for an event to be considered not controllable. For Tier 2 (≥25 mph sustained wind speed) events, source areas are presumed reasonably controlled if the control measures in place at the time of the event for all relevant anthropogenic source areas have been identified in a SIP that was approved by EPA within the past 5 years. If there is a SIP but it was approved more than 5 years prior to the event, some level of documentation is necessary to satisfy the requirement that the relevant sources were reasonably controlled. Otherwise, air agencies can submit a control analysis to demonstrate reasonable

---

[2] EPA defines a sustained wind speed as one where there is at least one full hour in which the hourly average was at or above the area-specific high wind threshold. EPA may also consider a sustained wind speed based on shorter averaging times on a case-by-case basis (EPA 2019b).

[3] Agencies can propose an alternate area-specific high wind threshold if it is shown to be more representative of the local or regional conditions. EPA guidance includes Ono (2006), which focuses on the Owens Lake region as an example of an area-specific calculation of a high wind threshold. Ono (2006) find a wind speed threshold (adjusted to 10-m height above ground level [AGL]) of 17 mph for the Keeler Dunes. Thus, local OVPA high wind thresholds may be slightly lower than the regional 25 mph criterion.

control measure implementation and enforcement. EPA (2019b, p. 11) notes that "the reasonableness of measures is case-specific and is evaluated based on information available at the time of the event" and that dust controls on anthropogenic sources are reasonable in any case where they "render the anthropogenic source as resistant to high winds as natural undisturbed lands in the area affected by the high wind event" (40 C.F.R. § 50.14 [b][5][v]). Additional analyses are required for Tier 3 (<25 mph sustained wind speed) events to justify that they are not reasonably controllable (EPA 2019b).

When considering the EER, it is important to understand that EPA was instructed by Congress to follow several principles in implementation of Section 319 of the CAA, including that "public health protection is the highest priority," and this applies even when attainment decisions are made with exceptions for high wind events. Section 319(b)(3)(A) states that "each state must take necessary measures to safeguard public health regardless of the source of the air pollution." Timely public notification is required when an "event occurs or is reasonably anticipated to occur which may result in the exceedance" of PM_{10}. All data are required to be included in a publicly accessible air quality database, and any exceptional event demonstration is subject to a public comment period of at least 30 days (40 C.F.R. § 50.14[c]).

APPLICABILITY OF EXCEPTIONAL EVENTS RULE FOR OFF-LAKE SOURCES IN THE OVPA

As discussed in Chapter 3, the OVPA is currently classified as being in "serious nonattainment." Since 2017, even excluding wildfire smoke, only two of the nine monitors—Lone Pine and Stanley—have consistently met the NAAQS criteria of no more than one exceedance per year averaged over 3 years (Table 5-1). Successful demonstrations for high wind exceedance events in the OVPA would remove those exceptional event PM_{10} exceedances in the determination of NAAQS compliance at that monitoring site. The District could reach regulatory attainment if all monitoring sites achieve the design value of less than one event per year, averaged over 3 years, as required by the NAAQS. Once regulatory attainment is achieved, the air agencies would be required to request redesignation and submit a Maintenance Plan SIP for approval. Monitoring and maintenance of all source areas would continue to ensure compliance with the NAAQS.

In 2017–2024, approximately 74 percent of exceedances have been attributed to sources other than "primarily on-lake sources," including wildfire smoke, regional events, and local off-lake sources (see Figure 3-4). However, the exceedance counts from recent years show that the on-lake exceedances alone at two monitors (Shell Cut, Dirty Socks) exceed the requirement of no more than one exceedance per year per monitor averaged over 3 years (see Figure 3-3). In 2021–2023 and 2022–2024, the Dirty Socks monitor recorded nine and eight exceedances, respectively, that have been attributed to on-lake sources (exceedance database, Chris Howard, GBUAPCD, personal communication, August 2024 and April 2025). These numbers are around triple those allowable for attainment. Some of the recent exceedances at Dirty Socks may have been sourced from an on-lake area (T1A-4A), where the Los Angeles Department of Water and Power completed contingency controls in 2023. Continued monitoring of the effectiveness of these on-lake controls will be helpful in evaluating on-lake control and the value of potential future application of the EER. Utilization of tools that calculate design values—like those produced by the South Coast Air Quality Management District—can help the District calculate if a reasoned use of the Exceptional Events Rule might be applicable and worth the investment of time and resources.

The panel was asked to "discuss the applicability of U.S. EPA's Exceptional Events [Rule] for excluding air quality monitoring data affected by dust from off-lake sources" during high wind events in the OVPA (Box 1-1, Figure 5-1). As described above, a key aspect of considering EER applicability for a given exceedance under high wind conditions is whether either of these conditions occurred:

1. the exceedance was a natural event, meaning that windblown PM_{10} was entirely from natural undisturbed lands in the area, where "human activity plays little or no direct causal role," or
2. reasonable control measures were applied at the time of the high wind event to control PM_{10} emissions from all anthropogenic sources.

TABLE 5-1 OVPA PM$_{10}$ Exceedances and Design Values by Site, Excluding Measurements with Exceptional Events Flags due to Wildfire

Site	2017	2018	2019	2020	2021	2022	2023	2024	2021–2023 Design Value	2022–2024 Design Value
Dirty Socks	9	2	2	3	8	12	6	8	8.8	8.8
Keeler	7	2	2	4	2	9	7	5	5.3 TEOM (POC 4) 6.2 Partisol (POC 6) 8.2 Partisol (POC 7)	6.7 TEOM (POC 4) 6.6 Partisol (POC 6) 8.2 Partisol (POC 7)
Lizard Tail	9	3	1	3	1	2	2	1	1.7	1.7
Lone Pine	1	1	0	1	1	1	1	0	1.0	0.7
Mill Site	2	1	1	3	0	4	2	3	2.0	3.0
North Beach	5	3	2	1	2	2	2	0	2.0	1.4
Olancha	4	2	0	1	0	6	4	1	3.4	3.7
Shell Cut	8	4	2	3	1	8	4	6	4.4	6.0
Stanley	1	0	0	1	0	0	0	1	0.0	0.3
Total Annual PM$_{10}$ Exceedances	46	18	10	20	15	44	28	25		

NOTES: Design values greater than one indicate that the area is not in attainment. PM$_{10}$ exceedances flagged by the District for exclusion under EPA's Exceptional Events Rule as a wildfire event are not shown. Design values are from EPA Air Quality System (AQS) as of April 2024 and April 2025 and exclude measurements with exceptional event flags due to wildfires. The Keeler site has three PM$_{10}$ monitors; the highest annual exceedance count among the three monitors is listed and the design value is shown for all three monitors (each monitor is denoted by a unique Parameter Occurrence Code [POC]). Design values in red are those that exceed the NAAQS criteria of no more than one exceedance per year averaged over 3 years. TEOM = tapered element oscillating microbalance.
SOURCES: Ann Logan, GBUAPCD, personal communication, May 2024, July 2024, April 2025.

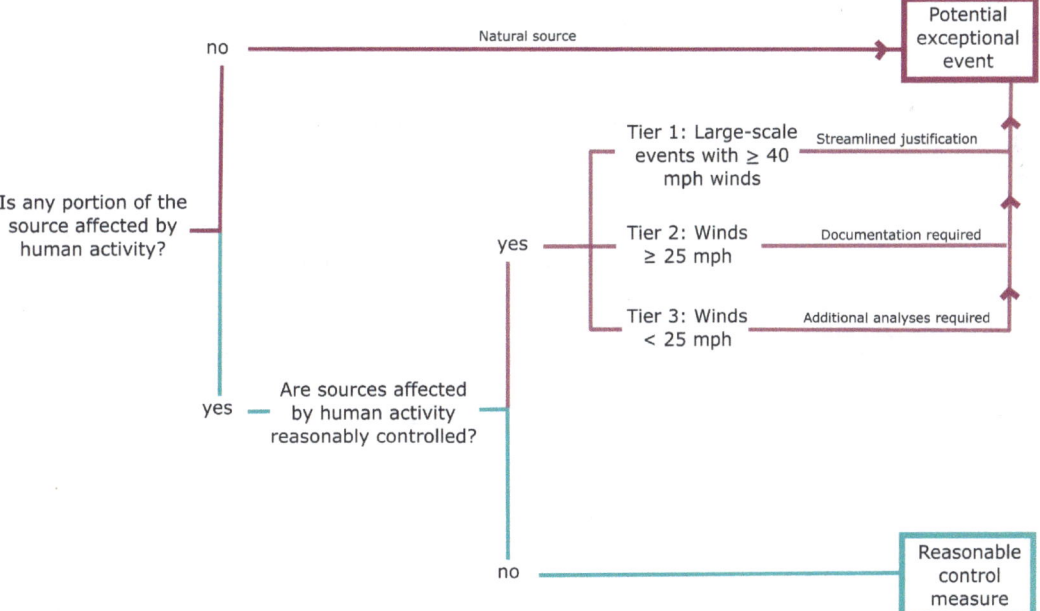

FIGURE 5-1 Decision tree to assist with evaluation of whether the U.S. EPA's Exceptional Events Rule is applicable to a high-wind dust event.
NOTE: A source is reasonably controlled if reasonable measures were implemented on anthropogenic sources at the time of the event.

Off-Lake Sources: Natural Event or Affected by Human Activity

The panel first evaluated what local off-lake PM_{10} exceedances might have been caused by a natural event, as defined by EPA. Based on the currently available evidence outlined in Chapter 4, there are a few PM_{10} source categories in the OVPA that the panel judges to be *entirely* natural, undisturbed lands. These include alluvial fans and many of the flood deposits, excluding those that have otherwise been affected by human activity, such as highway infrastructure. The panel also judges that the Swansea Dunes are likely natural, based on position and general dune movement over the past century, although this is based on limited information (see Chapter 4).

Other sources of PM_{10} emissions in the OVPA were deemed by the panel to be partially anthropogenic in origin. As described in Chapter 4, these include the Keeler Dunes, which were formed prior to lake drainage but were augmented by sediments from the dry lake surfaces that destabilized the vegetated dunes in a way that increased dust emission potential. Another likely source of partial anthropogenic origin are the flash-flood deposits south of Dirty Socks that were potentially enlarged in area by highway construction, which blocked the flow of floodwaters (see Figure 4-19).

As described in Chapter 4, the Olancha Dunes have somewhat expanded to the south in recent years, but it is unclear if this is related to drainage of the lake or recreation, including off-highway vehicle (OHV) use. Although no research has been completed on the impacts associated with decades of OHV activity at Olancha to PM_{10} exceedances, research at other sites shows clear associations between OHV activity, decreasing vegetation cover, and increased dust emissions (Cheung et al. 2021; Gillies et al. 2022; Groom et al. 2007; Walker et al. 2023). Chapter 4 includes recommendations for additional analysis to determine the effect of human activity on emissions from the Olancha Dunes. If these areas that are affected by human activity contribute to off-lake exceedances, a control measures analysis would need to be conducted, and reasonable controls would need to be implemented before exceptions under the EER could be considered.

Finally, there are several flood deposits described in Chapter 4 (e.g., Keeler, Centennial Wash) that were influenced by highway berms or a limited number of culverts, but it remains unclear whether the highway infrastructure affected PM_{10} emissions beyond what would be expected without that infrastructure in place. EER demonstrations for exceedances originating from flood deposits potentially influenced by the highway may benefit from additional analysis that demonstrates that PM_{10} emissions from these flood deposits have not been exacerbated by roadway infrastructure.

Reasonable Control Measures Applied

The panel also evaluated the "reasonably controllable" criterion for off-lake sources. EPA (2019b, p. 16) states that "an event involving windblown dust solely from natural undisturbed landscapes is considered a natural event and therefore not reasonably controllable." For sites affected by human activity, however, air agencies are required to demonstrate that an event is not reasonably controllable. As described previously in this chapter, an event is not reasonably controllable if reasonable measures were implemented on anthropogenic sources at the time of the event. It is beyond the charge to the panel to determine what controls are "reasonable measures" because "reasonable" includes policy judgments rather than purely scientific assessments. However, the panel notes that a thorough consideration of reasonable controls would consider resilience under the impact of climate change (e.g., planning for more intense flooding and prolonged drought). An assessment of reasonable controls could be part of a controls analysis that the EPA would require for the OVPA, as the latest approved SIP (GBUAPCD 2016) is over five years old (EPA 2019b). EPA (2019b) states that it will use the local list of reasonably available control measures (RACM) or best available control measures (BACMs), as applicable, as a reference point to review the reasonableness of new controls, and they note that more stringent controls may be reasonable for areas like the OVPA, which have frequent exceedances due to high winds.

Within the focus area of this study, the only off-lake source with current dust control measures is the Keeler Dunes, where emissions have been reduced by implementation of the approximately 170-acre Keeler Dunes Dust Control Project (see Chapter 6, Box 6-1), which was previously determined to be BACM (Federal Register vol 81, no. 238). A controls analysis would also be needed to determine if there are reasonable controls (Chapter 6) that could be implemented for exceedances from other sources determined to be anthropogenically influenced (e.g., flash flood drainage pooling caused by the highway).

The above evaluation for any given exceedance event requires consideration of wind speed using the EPA (2019b) three-tiered approach that determines the level of justification needed to assess if reasonable measures were implemented. Few off-lake exceedances (7 of 94) between 2017 and July 2024 in the OVPA have ≥40 mph sustained wind speed, which is one criterion for Tier 1 events (Figure 5-2). Nearly all off-lake exceedances met the sustained hourly wind speed of ≥25 mph for Tier 2 (Figure 5-2). However, the panel did not do a thorough investigation of a potential causal relationship between wind speed and PM_{10} emissions for every exceedance, which is required for an exceptional events demonstration (EPA 2019b). Only three exceedances between 2017 and July 2024 had sustained winds of <25 mph (Tier 3), and two of these had sustained windspeeds of over 24 mph.

PUBLIC HEALTH IMPACTS AND A REASONED USE OF THE EXCEPTIONAL EVENTS RULE

As described in Chapter 1, PM_{10} particles can penetrate the lungs and cause or worsen a variety of health problems like asthma, bronchitis, chronic obstructive pulmonary disease, and respiratory infections. Section 319 of the CAA indicates that "public health protection is the highest priority," and this applies even when attainment decisions are made with exceptions for high wind events. Section 319(b)(3)(A) states that "each state must take necessary measures to safeguard public health regardless of the source of the air pollution." However, Congress has recognized that it "may not be appropriate for the EPA to use certain monitoring data collected by the ambient air quality monitoring network and maintained in the EPA's AQS database in certain regulatory determinations" (40 C.F.R. § 52). Instead, Congress provided the statutory authority for the exclusion of data influenced by exceptional events (i.e., those events that are sourced from natural, undisturbed sources and anthropogenic sources that are not reasonably controllable). Exclusion of data from these PM_{10} sources may not improve the quality of polluted air in the OVPA, although the CAA requires timely information to alert the public whenever air quality is considered unhealthy. Exclusion of exceptional events provides a means for areas with high natural background levels of PM_{10} to attain the NAAQS, which can promote and encourage action on other reasonably controllable sources.

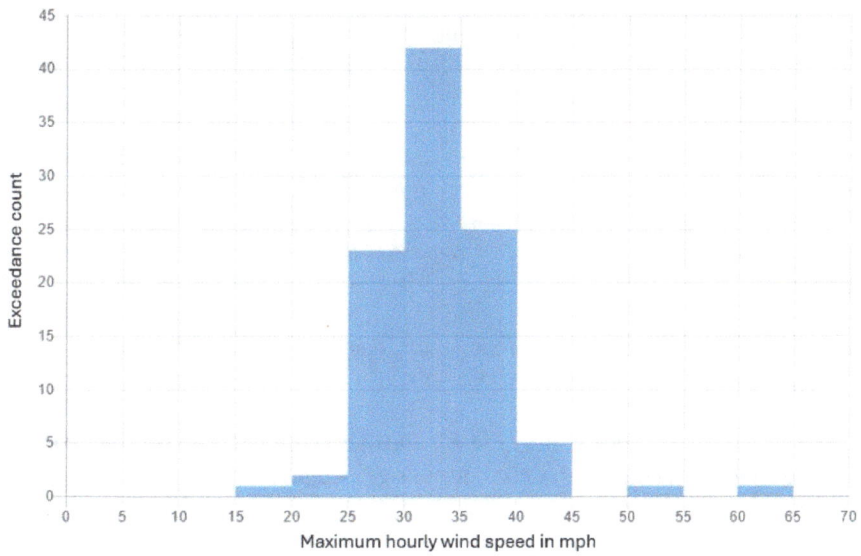

FIGURE 5-2 Off-lake exceedance counts and associated maximum hourly average wind speed, 2017–2024.
NOTES: The graph shows maximum 1-hour average wind speed from off-lake exceedance events between 2017 and July 2024. Wind speed is generally measured at a height of approximately 10 meters (GBUAPCD 2024b).
SOURCE: Data based on exceedance and wind speed data from GBUAPCD.

If PM_{10} emissions from natural or not reasonably controllable areas increase in the future due to climate change, the OVPA could experience worsening air quality (see Chapters 2 and 4). Under the Exceptional Events Rule, determinations about what is "natural" or "reasonably controllable" need to be made with care to prevent worsening air quality from anthropogenically influenced areas. Since changes in EPA's air quality regulations and exceptional events policies in 2007, which both increased stringency of air quality attainment and also included wildfires as natural events, there has been a marked increase in days that are submitted for exclusion of this data from regulatory determinations, and these are largely driven by wildfire and high-wind dust events (David et al. 2021). Reasoned use of the Exceptional Events Rule involves careful consideration of the anthropogenic drivers of PM_{10} emissions and the policy, health, and economic implications of controls. Acknowledging the long-term management approaches that enhance wildfire and dust emissions in otherwise "natural" areas as anthropogenic drivers of PM_{10} emission could incentivize local to regional efforts to proactively manage for ecosystem resilience in a changing climate and mitigate the effects of wind, fire, and storms on air quality (Clifford 2022; David et al. 2021; Richmond 2019). Such proactive management of natural lands can include prescribed fires to mitigate larger wildfires (EPA 2019a), changes in water management, and vegetation restoration.

CONCLUSIONS AND RECOMMENDATION

The Exceptional Events Rule for high wind dust events can be used to exclude unusual or naturally occurring exceedance event data from consideration in regulatorily significant decisions, as long as other public health protections are met. Nevertheless, the way the Exceptional Events Rule is interpreted and implemented has consequences for a region's air quality and associated health impacts. Based on current EPA guidance and the panel's evaluation of the origin and evolution of off-lake sources, some exceedances from off-lake sources in the OVPA may be considered natural events, as human activity played little or no direct causal role.

Conclusion 5-1: Most exceedances from channelized flow deposits, sheet flow deposits, and deposits impounded behind natural features appear to fit the Exceptional Events Rule criterion of natural events.

Some emissions from off-lake sources in the OVPA may be considered anthropogenically influenced. Since the OVPA has a State Implementation Plan that is over 5 years old, the Exceptional Events Rule requires an assessment of controls that could be applied to these anthropogenically influenced sources and the implementation of reasonable controls. It is beyond the charge to the panel to determine what controls are "reasonable" because that includes policy judgments rather than purely scientific assessments.

Conclusion 5-2: Emissions from the Keeler Dunes, the Olancha Dunes, and the highway-impounded flood deposit south of the Dirty Socks monitor have been affected by human activities resulting from draining of the lake, OHV recreation, and highway construction, respectively. Therefore, an assessment and potential implementation of reasonable controls would be required before an exceptional event demonstration is considered at these locations.

A thorough consideration of reasonable controls would consider resilience of the dust control measure under climate change. These considerations may include designing for more intense flooding and prolonged drought.

Recommendation 5-1: The District, California Air Resources Board, and the Environmental Protection Agency should consider resilience under climate change as part of its assessment of reasonable controls.

6

Informing Dust Control Decisions for Off-Lake Sources

Dust control measures (DCMs) on the Owens Lake bed and elsewhere have proven their effectiveness in reducing PM_{10} emissions (O'Brian 2021; NASEM 2020). Prior to the implementation of DCMs in 2000, Owens Lake was North America's largest single point source of PM_{10}, or particulate matter with an aerodynamic diameter of 10 micrometers or less (Reheis 1997). As detailed in Chapter 3, between 2001—the first year following the initiation of construction of the U.S. Environmental Protection Agency (EPA)-approved best available control measures (BACMs)—and 2023, significant declines in both the number of exceedance days and the average 24-hr PM_{10} exceedance values have occurred (see Table 3-1). While PM_{10} exceedances have steadily declined from on-lake sources since 2001, exceedances from off-lake sources have held relatively steady over time (Figures 3-6, 3-9, 3-12, 3-18, 3-22, and 3-24; Chapter 3). In the past 3 years, the number of exceedances at all monitors combined from off-lake sources has been greater than the number of exceedances attributed to on-lake sources (Figure 3-3). Chapter 3 identifies the many landforms within this area of the Owens Valley Planning Area (OVPA) that contribute to PM_{10} exceedances, including flood deposits, dunes, and alluvial fans.

In this chapter, the panel examines potential PM_{10} emission control strategies for sand dunes and associated sand sheets and flood deposits beyond the Owens Lake regulatory shoreline. The panel does not attempt to evaluate what controls are reasonable, given the costs and other impacts, as this is a policy decision that is dependent on more than just science. However, the panel recognizes that several of the BACMs approved for use on the lakebed that are discussed at length in NASEM (2020) may be inappropriate for off-lake conditions. For example, the shallow flooding BACM and its variations would not be feasible off-lake due to the greater depth to groundwater, permeability, and sloping topography. The following discussion focuses only on dust control measures suitable for off-lake sources.

It is important that the implementation of dust control measures also include monitoring to establish the reduction in PM_{10} emissions, as well as changes in soil texture, organic content, nutrients, moisture, and vegetation density. Vegetation density may be monitored remotely by aerial photography and by satellite sensing in the near future. Monitoring of soil conditions will require periodic sampling and analyses. Monitoring of dust emissions may be accomplished by remote ground-based cameras, remote sensing using satellite-mounted spectrometers, or in-situ monitoring (see Chapter 3 for discussion of low- to medium-cost PM_{10} monitors). This monitoring can be regularly analyzed to help assess the long-term effectiveness of different management approaches over space and time. It cannot be expected that dust management strategies will be effective in perpetuity, especially under a changing climate and in the face of extreme events (e.g., floods). If incorporated within an adaptive manage-

POTENTIAL DUST CONTROL MEASURES ON SAND SHEETS AND DUNES

There are several sand sheets and dune fields in the OVPA that are associated with PM_{10} exceedances, including Keeler and Olancha (Chapter 3). The only off-lake sand sheet or dune area with implemented DCMs is the Keeler Dunes Dust Control Project, which could serve as a potential model for future DCMs if they were deemed necessary.

Artificial Roughness

Artificial roughness elements placed on surfaces of sand sheets and active dunes increase aerodynamic roughness to reduce wind velocities at the surface and offer shelter that prevents sand cascading (Dong, Fryrear, and Gao 1999; Potter and Zobeck 1990). Roughness elements create downstream turbulence with eddy scales related to their size. Large roughness elements such as buildings can create eddies on the scale of several meters, while smaller roughness elements, such as surface clods and vegetation, tend to create much smaller eddies on the scales of millimeters to centimeters. The inter-molecular friction in a field of eddies dissipates the energy of the flow. Natural roughness elements that have been used to control dust at other sites include soil clods, straw checkerboard, standing and flat crop residues, baled crop residues, or woody residues of trees and shrubs (Dong, Fryrear, and Gao 1999; Qiu et al. 2004). Roughness elements constructed of engineered materials such as plastic and metal are termed engineered roughness elements. For the purposes of this report, artificial roughness as a dust control measure is divided into four types: solid natural, porous natural, solid engineered, and porous engineered.

Natural Roughness Elements

Solid natural roughness elements. The simplest natural solid roughness element is a straw bale. Straw is harvested and baled into rectangular or round bales as part of removal of crop residues from harvested land before seeding a subsequent crop. Straw bales are inexpensive ($3 to $6 per bale in the Midwest United States; Halopka 2022), typically have little feed value, and are high in lignin, adding to their longevity in the environment. Straw bales add roughness and reduce surface friction velocities of the wind. By simply placing rectangular straw bales on Keeler Dunes, the Great Basin Unified Air Pollution Control District (or District) reported an average reduction in sand flux of 85–94 percent (Box 6-1). However, the primary benefit of straw bales is that the bales reduce horizontal sand flux and create safe zones on the lee side of the bale that protect seedling shrubs from sandblasting. As the shrubs become established and grow larger, they augment the aerodynamic roughness provided by the bales, further reducing surface wind speeds, saltation, and dust emissions. Thus, the maximal benefit is provided by combining the straw bales with shrub planting, as described in Box 6-1 on the Keeler Dunes Dust Control Project. (Shrubs as a dust control measure are discussed later in this section.) Straw bales have also been used to control sand movement and dust emissions at the Oceano Dunes State Vehicular Recreation Area, where damage to dune vegetation from off-road vehicle use has resulted in destabilized dunes and enhanced dust emissions (Gillies and Lancaster 2013; Gillies, Furtak-Cole, and Etyemezian 2020). In this region, straw bales were shown to reduce shear stress within the roughness array and result in a decline of saltation downwind of the array by roughly 40 percent. Straw bales and other roughness elements including sand fences were only considered as temporary solutions for mitigating dust emissions at the Oceano Dunes and have since been replaced with permanent vegetation treatments (State of California Department of Parks and Recreation 2024).

Straw bales may enhance habitat by fostering the survival of seedlings. The straw offers little in the way of food for migratory wildlife, but the mechanical integrity of the bale does allow for burrowing animals to create hibernation and nesting sites in the unstable sand. This could provide ecological benefits as animal waste products and organisms carried on the feet of the wildlife would enrich the sand and enhance its ability to support plant growth. While the placement of straw bales offers no cultural value to the site, these projects may be somewhat beneficial to the local economy if the straw is grown on local land or installed by local community members.

BOX 6-1
Keeler Dunes Dust Control Project

As part of the 2013 Stipulated Order of Abatement and Settlement Agreement, The Los Angeles Department of Water and Power (LADWP) was released from liability for dust emissions from Keeler Dunes, Swansea Dunes, and Olancha Dunes after contributing $10 million for a public benefit contribution to the District (GBUAPCD 2016). The District used these resources to initiate the Keeler Dunes Dust Control Project, which used straw bales as a temporary dust control measure to protect new plantings of four native, locally adapted shrub species—*Sarcobatus vermiculatus, Atriplex polycarpa, Atriplex parryi,* and *Sueada moquinii (nigra*; Figure 6-1). The overall goal of the project is to reestablish a stable vegetated dune system. Shrub seedlings were grown out at nurseries from locally collected seed and transplanted into the protected zones near the straw bales. Between 2015–2024, a total of 238,000 seedlings were planted within the 170-acre project footprint, creating approximately 140 acres of dust control measures. The District also built an above-ground water distribution system to facilitate hand watering during planting and establishment. Water use over this 10-year period averaged 2.3 acre/feet per year. Between 2019 and 2023, there was 14 percent shrub cover with an additional 10 percent straw bale cover within the project footprint. Plant monitoring in June 2023 found that 62 percent of the shrubs at the site were planted and 35 percent were natural recruits. Roughly equal proportions of the four species were observed, and 83 percent were more than 3 years in age. Based on the calculated average sand flux values during the 3 years before the start of the project (2011–2013) and during the last 3 years of available data (2021–2023), three sites displayed similar extents of sand flux

FIGURE 6-1 Straw bales on Keeler Dunes in May 2024.
SOURCE: Ian Walker, panel member.

continued

BOX 6-1 Continued

reduction (92–94 percent or factors of 13–16 reduction) whereas one site showed an 85 percent or a factor of 7 reduction. However, all four sites have a statistically significant decreasing trend in the annual sand flux value (Mann-Kendal test, p-value <0.05) during 2011–2023. PM_{10} exceedances attributed to the Keeler Dunes region also decreased from 2015 to 2021 (Figure 3-18), although exceedances increased in 2022 and 2023, potentially due to recent flash flood deposits. As they continue to grow, these shrubs will further increase the dust reduction efficiency by reducing surface wind velocities, filtering airborne sediment in their canopies, and stabilizing the near surface sand with litter and deeper soil horizons with root structures.

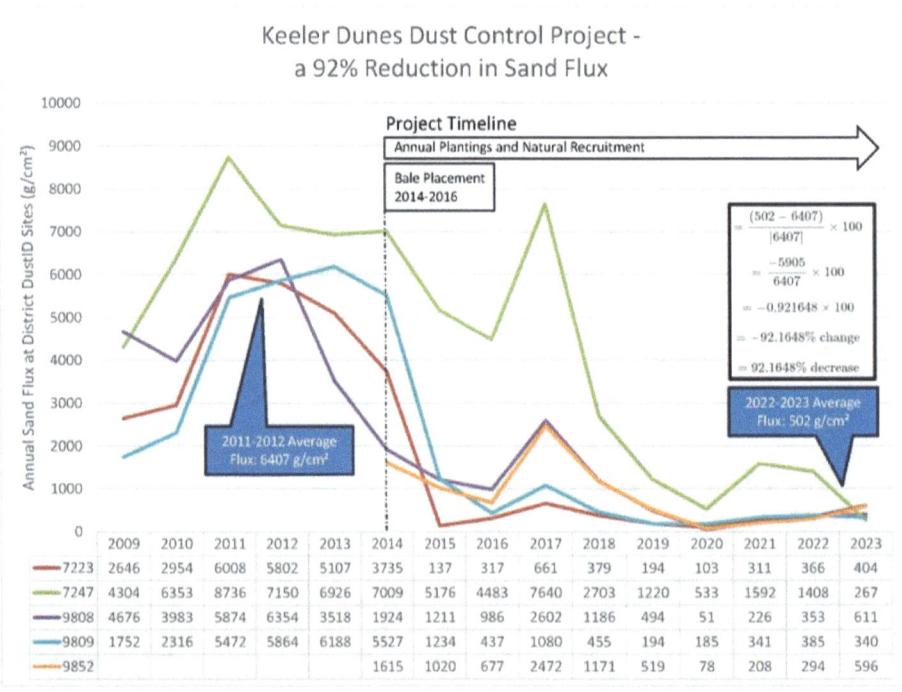

FIGURE 6-2 Keeler Dunes Dust Control Project led to an 85–94 percent reduction in sand flux.
NOTES: The different colored lines (7723, 7247, 9808, 9809, and 9852) represent different monitoring sites within Keeler Dunes.
SOURCE: Ann Logan, GBUAPCD, July 2024, personal communication.

Porous natural roughness elements. Porous natural roughness elements have not been tested in the OVPA but may offer sand drift and dust emission control. Porous natural elements include the woody skeletons of trees and shrubs laid on the surface and straw inserted into the sediment surface vertically in a checkerboard pattern (Figure 6-3). The woody skeletons may offer promise in off-lake areas with cultural and historical sensitivity since they do not require modification to the ground surface. However, some of the larger woody skeletons may require large equipment to transport them onto the site.

Natural porous surface roughness created with plant matter (e.g., straw) placed in checkerboard patterns (Figure 6-3) have been shown effective at controlling sand movement and dust emissions (Bo et al. 2015; Li et al. 2006; Wang, Qu, and Niu 2020; Zhang et al. 2004; Zhang et al. 2018; Zhao et al. 2008) and have been used for this purpose for over a half century in northwestern China (Qiu et al. 2004). The straw checkerboards are typically established in 1-m^2 grids with an optimal height of straw 20 cm above the ground (Bo et al. 2015; Qiu et al. 2004) These straw checkerboards reduce wind speed near the surface to steady-state velocities less than threshold—which is the wind speed value necessary to initiate soil erosion and dust emissions within about 66 ft (20 m) of the upwind edge (Xu et al. 2018). Narrow bands have been shown to be as effective as solid straw checkerboard arrays of much greater areal coverage (Bo et al. 2015). This design results in a reduction of horizontal sand flux by as much as 99.5 percent (Qiu et al. 2004). Straw checkerboards are adaptable to sloping and undulating land, making this a flexible method for complex landscapes (Dehkordi et al. 2023; Lihui et al. 2015).

Straw checkerboards also trap dust and encourage soil formation and the colonization of the stabilized sand by vegetation (Li et al. 2006). Soil components and plant macronutrients such as nitrogen, phosphorus, and potassium carried on dust have increased in straw-checkerboard-bounded soil (Dehkordi et al. 2022) and a 10-year study demonstrated increases in silt, clay, organic matter, water availability, nutrients, and pH (Wang and Wang 2019). Li et al. (2020) report that microbially induced carbonate precipitation from calcium ions carried on dust lessen the temporal limitations of straw checkerboards by accelerating sand fixation through crust formation, and this stabilization results in accelerated vegetation recovery. Narrow bands of vegetation watering might be highly beneficial immediately upwind of large areas with straw checkerboards to limit burial of the windward squares of the checkerboard and to provide an immediately adjacent source of plant litter and seeds for sand colonization and stabilization by native vegetation.

Straw checkerboards are historically built with manual labor in China, and a single laborer can create about 200 m^2 per day. Recently, machines have been developed that increase the efficiency of installation (Figure 6-3). In China, straw checkerboards are functional for up to 4 years, but the warm and dry season in California may

A   B

 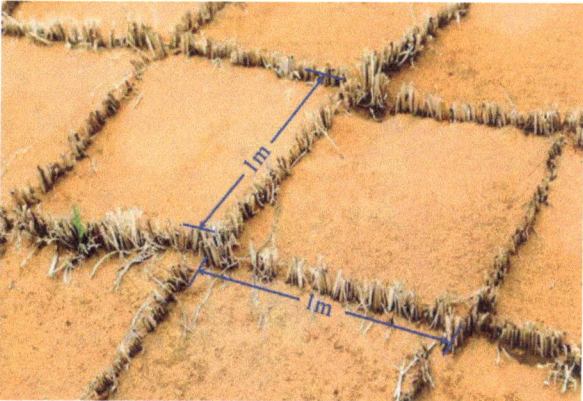

FIGURE 6-3 A) Gasoline powered machines insert straw to form straw checkerboard barriers. B) Sand dunes in western China covered by straw checkerboard.
SOURCE: Photos courtesy of Guoming Zhang of Beijing Normal University.

result in less microbial degradation and longer function, similar to what has been noted with the straw bales. If more time is needed to protect establishing vegetation, new or partial straw checkerboards could be implemented where needed.

The ecological value of the straw checkerboards is primarily related to potential soil quality improvements and enhanced plant colonization, which in turn improve habitat for local wildlife. Straw checkerboard barriers provide minimal cultural value but may provide machine operator jobs locally and opportunities for local farmers to sell wheat or barley straw.

Engineered Roughness Elements

Solid engineered roughness. Solid engineered roughness elements are built from materials such as metal, concrete, or plastic and placed in an array to reduce near-surface wind speed through added friction. For example, Gillies (2017; 2018b) investigated dust control effects of a staggered array of plastic boxes (2.4 ft x 1.5 ft x 1.2 ft [0.725 m x 0.45 m x 0.38 m]; Figure 6-4). Maximum sand control efficiency of 90 percent was achieved approximately 200 ft (60 m) from the upwind edge of the array. The authors maintained that the control efficiency and distance to achievement is a function of the density and distribution of the roughness elements (Gillies et al. 2017, 2018b).

Although solid engineered roughness elements could be arranged on a sand dune or sand sheet as needed to reduce PM_{10} emissions, they do not blend well with the natural landscape, have no intrinsic value to wildlife, and depending on the material from which they are constructed, are susceptible to photodegradation and other forms of weathering. For instance, polyethylene would degrade in approximately 1 year and photodegradation of the plastic could lead to release of microplastics into the environment, while concrete might still be serviceable after 50 years.

Porous engineered roughness elements. Engineered roughness elements may be constructed to provide specific

FIGURE 6-4 Weighted plastic totes used as engineered solid roughness objects on Owens Lake.
SOURCE: Grace Holder, GBUAPCD, personal communication, August 2019.

amounts of porosity in order to optimize the sheltering and aerodynamic roughness effects. These include porous structures and sand fences constructed of porous polymer sheets.

Porous engineered roughness elements have been tested as a dust control method at Mono Lake, California. Gillies, Etyemezian, and Nickolich (2018a) found that the porous engineered roughness elements achieved maximum control efficiency at approximately half the distance from the upwind edge of the array when compared with solid roughness elements and resulted in less near-element scour. Like the solid engineered elements, porous engineered roughness elements offer low aesthetic and wildlife value and have similar longevity.

Another type of porous engineered roughness element used at Owens Lake that could be considered for the off-lake area is sand control fencing made of various composite and woven materials (Figure 6-5). Sand control fences work by offering a physical barrier to stop saltating sand cascades and by reducing the wind velocity at the surface in front of and behind the fence. The relative effectiveness of such fences at reducing wind velocity is a function of the material and density. For instance, barriers of vertically placed parallel plastic pipe with an optical density of 12.5 percent (2.8 cm diameter pipes placed 18.7 cm apart) only reduced wind velocities by an average of 4.3 percent in the 33 ft (10 m) upwind and 99 ft (30 m) downwind region. In contrast, a barrier of vertically placed parallel plastic pipe with 75 percent optical density (1.1 in [2.8 cm] diameter pipes placed 0.3 in [0.9 cm] apart) reduced wind velocity by an average of 32.5 percent (Bilbro and Stout 1999). Maximum wind velocity reduction is found near the barrier, and the barrier effect diminishes with distance from the barrier. The height of the barrier also influences the effectiveness of sand control. Computational fluid dynamics programs have been used to design sand control fence networks (Xin et al. 2021).

Sand control fences are most effective when they contact the ground surface, but sediment may need to be removed periodically. In the case of more porous sand control fences, sand removal may be necessary on both sides of the fence as eddies in the lee of the elements result in sand deposition. Sand fences provide no habitat value, and land-based wildlife may need occasional gaps in the fence to provide for migration. The process of

FIGURE 6-5 Sand fence at Owens Lake.
SOURCE: LADWP. Photograph by Mark Schaaf, Air Sciences.

installing sand fences could damage cultural resources in the area. Sand control fences may favor vegetation by trapping finer particles, plant litter, and seeds in wind transport (Zhang et al. 2007).

Shrubs as Perennial Vegetative Cover

For dune and sand sheet settings, vegetative cover is widely considered the primary surface protection from the erosive force of winds. Shrub communities are commonly found in off-lake areas surrounding Owens Lake (Chapter 2). Vegetation reduces aeolian transport—and therefore PM_{10} emission—by mechanisms that are similar to those described previously with regard to porous roughness elements. Shrub communities reduce PM_{10} emissions by reducing the wind velocities near the surface, capturing airborne sediments within their canopies, and further stabilizing the surface with their root mass and shed biomass. Shrubs and larger grasses (as opposed to flat or very low vegetation) also have a wake area with reduced surface shear stress in their lee, which protects large areas from emission, more efficiently dissipates momentum from the wind, and captures more (and higher) airborne material (Raupach et al. 2001). Although not as effective as crop residues or native grass cover, shrubs affect the wind fields and, in desert environments, may offer the only natural, self-sustaining protection of the soil surface from wind. Shrubs also serve as rigid structures into which moving sediment may become trapped.

A shrub-based dust control measure, with shrub densities of 2 per ft^2 (approximately 0.2 per m^2),[1] or roughly 10 percent cover, are expected to have >85 percent control efficiency within 82 ft (25 m) of the edge of the shrub area (Li et al. 2013; Mayaud, Bailey, and Wiggs 2017). It has not been tested whether >95 or 99 percent control efficiency can be attained, although preliminary simulations by the panel assuming porous vegetation estimate that >20 percent shrub cover may be required (see Box 6-2). Due to edge effects, relatively large areas (>10 hectares) are necessary to ensure that a majority of the area is within target control efficiencies.

Shrub-based dust control has the potential to provide habitat for native and migratory species. Of the habitats found at Owens Lake, shrublands support the most diverse species of lizards, snakes, birds, and mammals (LADWP, 2010). Shrubs may be more accepted as a way to reduce aeolian transport in culturally sensitive areas, though the impact of infrastructure for delivering water is a potential issue. Shrubs also have positive aesthetic value.

Two approaches for increasing vegetation density in sand dunes and sand sheets include shrub planting and using connectivity modifiers to support vegetation growth between existing shrubs.

Planting Shrubs

In dunes that have been affected by human activity, shrub planting efforts can provide a natural and sustainable approach to reduce PM_{10} emissions. Initially, shrubs planted at the correct density will not be able to provide this control efficiency until they grow larger because nursery shrubs require 5 to 10 years to grow into mature plantings, depending on the species. Fertilization may help increase the rate of growth, but the resulting fast growth may also reduce the amount of lignin in plant cells, which could impact the ability of the shrubs to survive sandblast injury. Most natives are adapted to low nutrient availability, and thus if fertilizer is used, it may only be needed sparingly. Additional temporary dust control will likely also be needed to provide near-term PM_{10} reduction. Densities that are too low have the potential to become unsustainable, as pedestaling and abrasion of shrubs by moving sand can cause dieback and mortality (Okin, Gillette, and Herrick 2006). Additional analysis would be necessary to determine the shrub density and species composition necessary to achieve the desired levels of PM_{10} control in off-lake dunes and sand sheets that are still contributing to exceedances (Box 6-2).

The planting of shrubs and their subsequent reproduction and colonization in the Keeler Dunes Dust Control Project (Box 6-1) is a great success story. The introduction of straw bales as temporary natural solid roughness elements in the active sand sheets and foredunes of the Keeler dune field allowed for protection of shrub seed-

---

[1] This density is assumed to estimate control efficiency from the Keeler Dune data, based on the fact that two shrubs are roughly equivalent to one hay bale. Shrubs are roughly the same height as the approximately 15-in (38-cm) high bales but approximately one-half the width of the 44-in [112-cm] wide bales. Thus, shrubs have a profile area of approximately one-half of the bales used on Keeler Dunes. See equation two in Gillies et al. (2015).

BOX 6-2
Modeling Shrub Density Requirements

Simple modeling in NASEM (2020) demonstrates the amount of vegetation cover that may be needed for 99 percent control efficiency. That analysis uses the Okin (2008) model of shear stress partitioning and 5-minute winds from North Beach. Using simple assumptions about vegetation size (1.6 ft [0.5 m] diameter) and aspect ratio (1), the model predicts that 25 percent shrub cover would provide approximately 99 percent control efficiency using the parameters in Li et al. (2013) for the Shao, Raupach, and Findlater (1993) saltation flux equation. Using more conservative parameters from Mayaud, Bailey, and Wiggs (2017), the model predicts that 38 percent shrub cover is required for 99 percent control efficiency. To obtain 95 percent control efficiency, 13 and 18 percent cover are required for the Li et al. (2013) and Mayaud, Bailey, and Wiggs (2017) parameters, respectively (Figure 6-6). Biological constraints on the sustainable densities of rainfed shrubs would need to be considered when establishing expectations for controls. Typical existing shrub communities on the basin floor have shrub covers of approximately 23 percent (Figure 6-6). Thus, whether solely rain-dependent communities can be established on dunes or sand sheets with densities sufficient to obtain 99 percent control efficiency is an outstanding question. It appears possible, however, that a rain-dependent shrub community may be able to provide at least 95 percent control efficiency if vegetation cover exceeds 20 percent.

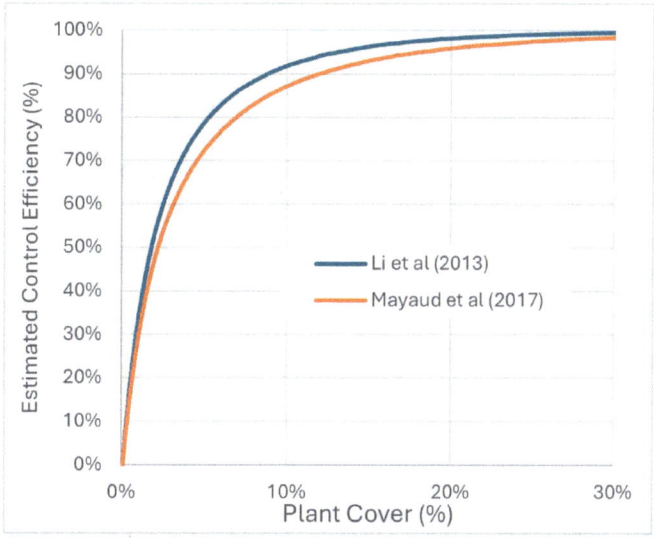

FIGURE 6-6 Estimated control efficiency using two sets of parameters in the Okin (2008) model using 5-minute winds from North Beach, assuming gamma distribution of plant spacing and plant height = plant diameter = 20 in [50 cm].
SOURCE: NASEM (2020).

lings from sandblast injury. A limited water distribution system combined with hand watering provided irrigation (approximately 0.1 ft/yr) for the seedlings until their root systems were sufficiently established to maintain growth. After shrubs have been established and have grown to target sizes, watering is expected to be tapered to zero, because the shrubs will be able to survive on local rainfall. In the last few growing seasons, the planted shrubs have been reproducing, and their offspring are colonizing the open spaces where control was previously lacking (Grace Holder, GBUAPCD, personal communication, May 2024).

Connectivity Modifiers

Desertification has been attributed to changes in ecosystem connectivity (Okin et al. 2009). In the last couple decades, shrubland plant communities have been replacing grasslands in semiarid and arid landscapes. Although the reasons are not entirely clear, changes in precipitation patterns due to climate change favor deep rooted shrubs over more shallow rooted grasses (Li et al. 2022). By increasing the distance between growing plants, more bare soil is available to be eroded by wind, and the topsoil is less likely to be deposited in shrub-based fertility islands, making plant establishment in the nutrient-deficient soil between the shrubs less likely. In addition to the loss of nutrients in the eroded topsoil, organic carbon is also lost and preferentially deposited in the shrub canopies and the soil directly below, making water less available between rain events in the bare interspaces. Thus, in the absence of factors that would reverse the state changes in connectivity, there is little chance of reversal (Bestelmeyer et al. 2015).

In an effort of modify the scale of connectivity in degrading drylands and sand sheets, researchers have used porous engineered structures as connectivity modifiers or ConMods (Okin et al. 2015; Rachal et al. 2015). ConMods are constructed of two pieces of 0.2 in (6 mm) mesh galvanized hardware cloth—both 0.8 ft (20 cm) tall by 2.2 ft (55 cm) long—that are attached to steel rods at each end and arranged crossing perpendicularly in the center of each piece to form a cross with a fifth pin anchoring the intersection (Figure 6-7; Rachal et al. 2015). ConMods are installed flush with the soil surface and may be placed on the landscape at user-selected spacings to trap seeds and litter that would otherwise only be trapped by a downwind shrub canopy. ConMods have been proven effective at trapping litter and seeds and helping vegetation establish in the formerly bare inter-shrub soil surfaces (Okin et al. 2015; Peters et al. 2020), effectively promoting degradation state reversal and increasing connectivity in plant cover, protecting the soil from the erosive forces of wind. Although engineered and offering no direct benefit to wildlife, they are small and not highly noticeable on the landscape from a distance. More work needs to be done with ConMods in order to determine the optimal density of their installation. They are immediately effective, but the effectiveness increases as litter is trapped and vegetation starts to grow. They may be removed at some future time, but care is needed not to damage the plants that have established. It may also be possible to build ConMods with biodegradable materials such as burlap and wooden dowels.

Limitations on Recreational Activity

Research at the Oceano Dunes State Vehicular Recreation Area and the Imperial Sand Dunes Recreation Area shows clear associations between off-highway vehicle (OHV) activity, decreasing vegetation cover, and increased dust emissions (Cheung et al. 2021; Groom et al. 2007; Walker et al. 2023). Following implementation of over 400 acres of dust control treatments at Oceano Dunes, including restored vegetation cover, there have been appreciable declines in PM_{10} emissions measured at receptor sites downwind of the mitigation treatments (State of California Department of Parks and Recreation 2024). In the Owens Valley, the Bureau of Land Management (BLM) manages an OHV recreation and primitive camping area of approximately 400 acres (1.6 km^2) or 36 percent of the total area of the Olancha Dunes, which has been operational for decades. Limiting recreational activity in areas that are highly susceptible to PM_{10} emission is one potential dust control measure at the Olancha Dunes.

FIGURE 6-7 A connectivity modifier (ConMod) stabilizing sand at the U.S. Department of Agriculture–Agricultural Research Service (USDA-ARS) Jornada Experimental Range.
SOURCE: Photo courtesy of Michael Fischella.

POTENTIAL DUST CONTROL ON FLOOD DEPOSITS

As described in Chapter 2, episodic floodwaters following rain events carry fine-grained sediments that deposit in the off-lake areas. As these flood deposits dry, the clay particles at the surface are highly vulnerable to erosion from the wind, and even armored clay surfaces can be abraded when bombarded with sand. Although flood deposits are recurring events that regenerate emissive sediment surfaces, dust control measures are available that could be used to reduce PM_{10} emissions, if deemed reasonable, as described below.

Cover by Gravel or Cobbles

Gravel coverage has been proven effective at controlling dust emissions on the lakebed where gravel (>0.4-in [1.25 cm] diameter) is laid in a 4-in (10 cm) layer over bare ground (GBUAPCD 2013) or in a 2-in (5 cm) layer over a geotextile (GBUAPCD 2013). The gravel not only shields the erodible soil surface from wind but also functions to limit surface evaporation. Gravel coverage can be used on off-lake flood deposits with complex topographies. NASEM (2020) also recommended consideration of cobbles, which are larger than gravel and have a greater capacity for capture and storage of windborne material compared to the more-uniform gravel. The advantage of gravel and cobble coverage is that these dust control measures require no water and have a control efficiency of >99 percent if not buried by encroaching sand movement. While gravel and cobbles are probably not practical for the sand sheet or dune areas due to the ease with which saltating sand can bury them and render them useless to control sand movement, they could serve to bury the fine clays of flood deposits and prevent future PM_{10} emissions.

The use of these sterile geological materials is often regarded as unsightly and offers little or no value to wildlife habitats other than potential protection of small rodents and insects. The machinery used to transport and spread the gravel could damage both natural and cultural resources in the area. Additionally, gravel and cobbles would likely only serve to reduce PM_{10} emissions until the next flood event.

Vegetation Establishment

Fine-grained flood sediments are typically associated with higher nutrient levels and water holding capacities than the surrounding coarse-grained deposits on sand sheets and alluvial fans. Thus, flood deposits might be suitable for vegetation establishment as a means to reduce dust emissions. This plan may require manipulation of the flood deposits by covering them with unlined layers of gravel or cobbles to allow for rapid infiltration of rainfall while limiting surface evaporation, creating an augmented environment for shrub establishment. Due to the depth of gravel in the gravel blankets, shrubs would need to be nursery-grown and transplanted to allow canopy to emerge above the gravel blanket while having roots in the finer sediments. While this vegetation would provide more resilient stabilization than just gravel and cobbles, the shrubs would need to be resistant to temporal flooding episodes. There may be native vegetation along streams and near shorelines of terminal lakes in this region that have such adaptations. For example, *Sacrobatus* has been shown to be tolerant of flooding conditions for up to 40 days (Ganskopp 1986), but vegetation is also susceptible to death through scour and uprooting during extreme storms. Longevity of this dust control measure would depend on the interval between floods, and barring changes to infrastructure or local topography, measures would potentially need to be renewed following a major flood event. Improved resilience of shrubs in flood deposit areas could be promoted through modifications of existing infrastructure that concentrates stormwater flows.

Modified Highway-Associated Floodwater Infrastructure

As noted in Chapters 3 and 4, one off-lake source of PM_{10} emissions to Dirty Socks appears to be a pooled flood deposit next to Highway 190 (Figure 4-19). Future extended flooding in this area following large precipitation events could be minimized by the construction of distributed culverts where the floodwaters are currently retained by the highway.

The California Department of Transportation (Caltrans) also built several chevron-shaped berms to protect Highways 136 and 190 by diverting the braided flow into channels for passage through culverts. The panel was unable to determine to what extent these berms affected PM_{10} emissions and exceedances compared to what would have occurred from flood deposits in the absence of this infrastructure (see Chapter 4). For instance, the 2022 image in Figure 4-18 shows the water flows during the high-energy storm event associated with the remnants of Hurricane Kay in 2022, when water flowed and pooled against the roadway, ultimately overtopping the road and damaging the roadbed. Instead of using berms to concentrate the flow to pass under the highway in a few culverts, removing the berms and creating many smaller culverts would facilitate passage of floodwater over a wider area, facilitating patterns of flow and infiltration in the off-lake area more similar to what it would have been had Highway 190 not been built. Additional culverts would also decrease the likelihood that water pools or flows behind (upstream of) the road creating potentially emissive flood deposits. The cost of reconstructing the highways with additional culverts and benefits of such modifications are uncertain. Once completed, however, there would probably be minimal annual maintenance costs.

Upslope Runoff Management

An alternative to modifications of highway features that concentrate flood flows would be upslope runoff management to encourage water spreading. Berms placed perpendicular to the flow tend to spread the flow laterally along the slope, enlarging the surface area available for infiltration and removing energy from the flood flows. Such berms have been used to heal eroded land, prevent gully formation, and enhance water harvesting for vegetation. Water harvesting is a relatively mature science used in arid regions globally (Boers and Ben-Asher 1982; FAO 1991). Water harvesting techniques often involve creating surface contours for diverting water to trees and shrubs, surface dimpling to enhance local storage and infiltration, and constructing permeable dams and other stormwater spreading features (Figure 6-8) to control water flow, offering flexibility in either spreading or concentrating the surface water flow as necessary (Mekdaschi and Liniger 2013). Increased infiltration on the fan itself would, with the exception of the most intense events, possibly prevent damage to on-lake dust control measures. In the Salton Sea Management Project, stormwater is spread laterally across the landscape to enhance infiltration and support vegetative growth for dust control and to leach salts from the surface soil. These approaches include using rock

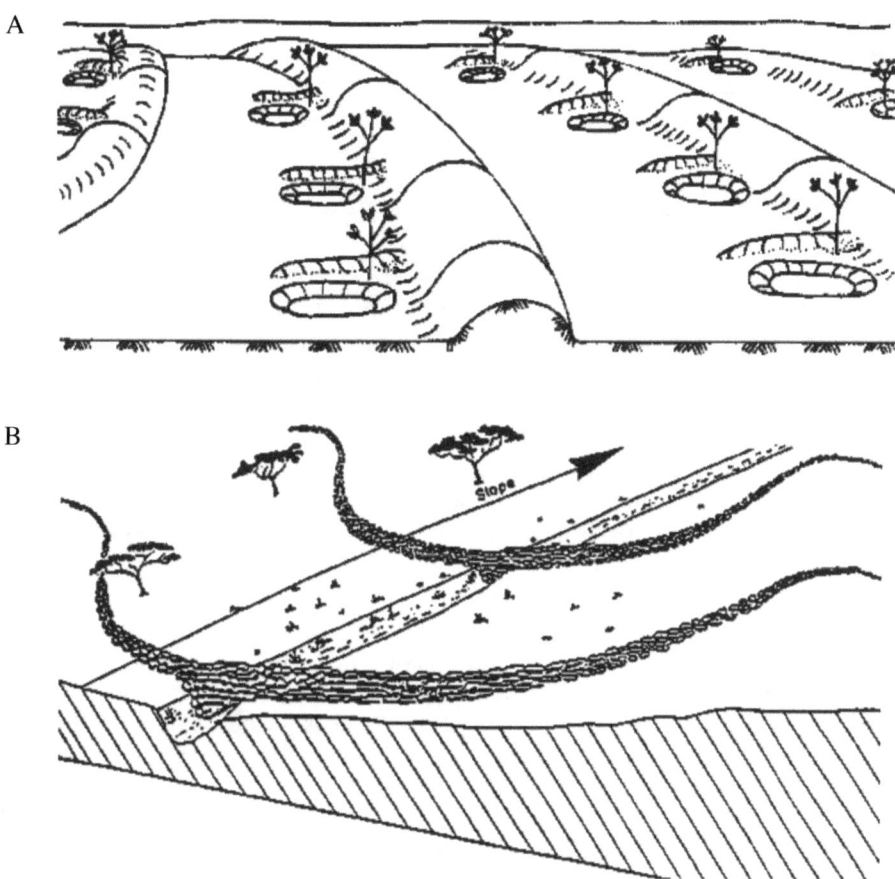

FIGURE 6-8 A) Water control bunds installed on the landscape to trap water at the downhill portion of the open area and uphill side of the bund, forcing it to spread laterally and enhance precipitation capture efficiency of vegetation. B) Permeable rock dams used to spread runoff waters for improved water availability to trees and shrubs while reducing scouring power of runoff in gullies. The structures are long rock walls that can be scaled (using larger rocks) to withstand expected runoff pulses feeding present flood deposits. Such structures assist with supplying and establishing vegetation while controlling patterns of fine deposits.
SOURCE: FAO (1991).

weirs on swales and berms and constructing ditches along contours and perpendicular to creeks (California Natural Resources Agency 2024).

Historically, tribes engaged in extensive spreading of water across the valley floor north of the Owens Lake using irrigation ditches to promote the growth of food crops and other plants important for cultural reasons, including meadow and riparian species (Lawton et al. 1976). This did not occur in the immediate proximity around Owens Lake, despite the presence of creeks that could support these practices, presumably because the lake already provided abundant food (Lawton et al. 1976), but this practice could be a promising approach to promote vegetation cover to control dust while providing important habitat. A notable concern with this approach is the initial disturbance of the land surface, typically accomplished by large machinery, which could impact cultural resources.

OTHER POTENTIAL DUST CONTROL MEASURES

A number of novel techniques have been employed in China to control sand movement and dust emissions. These include producing seedlings of drought-adapted native trees and shrubs and planting them in small basins imprinted to deliver rainfall to the base of the seedling, as well as spreading plastic mulch over irrigated farm fields (Lyu et al. 2020).

In other areas of Asia plagued with sand and dust storms, the biopolymer Acacia gum has been used in the laboratory to stabilize loose sand (Dagliya and Satyam 2024) and the authors also report on similar studies utilizing other natural plant products to bond sand grains together. Other non-plant-based sprays and surface applications have been investigated including emulsions of asphalt, polyvinyl alcohol, styrene-butadiene, latex, and water-miscible resins applied in water dilutions at the rate of 3,785 liters per hectare (Armbrust 1977). The author also reports that feedlot manure applied at the rate of 31.8 tons per hectare to the surface, or 52.3 tons per hectare plowed into the substrate will also effectively control wind erosion. Tatarko, Trujillo, and Schipansk (2019) found that such treatments are often costly and temporary due to degradation from irrigation, sunlight, rainfall, tillage, or abrasion from adjacent untreated areas. In addition, many of these treatments are phytotoxic and not compatible with all soil types (Presley and Tatarko 2009). The Salton Sea Management Project evaluated multiple soil stabilizers and concluded that most would be temporary because they are soluble in water, but that soil binders may be useful in limited locations. For example, binders can create a hardened surface that can withstand heavy vehicle traffic and can be chemically stable in desert conditions with various sized particles (Environmental Science Associates 2022). Although soil stabilizers on their own may not be promising, they can be an extremely useful tool for temporary stabilization of soils for biocrust restoration (Faist 2020a).

Biocrusts have been observed in the "barren" areas of Owens Lake Playa (LADWP and Ecosystem Sciences 2010). Biocrusts are a consortium of organisms (e.g., cyanobacteria, green algae, bacteria, microfungi, lichens, mosses) that live at or near the soil surface and stabilize soil directly through their above-surface cover (e.g., mosses and lichens), their filaments and hyphae, and their secretions that aggregate soil particles (Belnap 2006; Rodriguez-Caballero et al. 2022). Many types of biocrust are relatively resistant to water and wind erosion but can be very vulnerable to burial or compression (e.g., footsteps, vehicles; Faist et al. 2020a). In the Mojave Desert, which is adjacent to Owens Valley, biocrust cover is greatest on 20–7,000-year-old surfaces and negligible on very young surfaces (e.g., active washes and recent sediment deposits). But even moderately active sand sheets can host cyanobacterial crusts, suggesting they are tolerant of some levels of burial and sand deposition (Bowker et al. 2016).

Although biocrust restoration has previously been viewed as futile as a dust control measure at the Owens Lake bed because it can be vulnerable to burial and was assumed to have a very slow natural recovery rate (NASEM 2020), recent work has demonstrated that biocrust recovery is possible in years to decades, in some instances. Environmental improvements such as soil stabilization, decreased ultraviolet (UV) radiation, and micro-irrigation can assist with biocrust recovery (Antoninka et al. 2020b; Fick et al. 2020; Zhou et al. 2020). In areas with unstable soil, soil stabilizers can be effective in short-term surface stabilization, facilitating the establishment of biocrusts, which can provide long-term surface stabilization in many drylands (Antoninka et al. 2020b; Faist 2020a). In biocrust restoration, preferred soil stabilizing agents are bio- or UV-degradable. Polyacrylamides and plant-based stabilizers (e.g., psyllium, jute [a woven cloth from plant-based materials]) have been shown to be effective (Faist 2020a; Fick et al. 2020). However, it may be important to repeat stabilization treatments for multiple years until biocrust is established (Faist et al. 2020b). In addition, straw checkerboards (discussed above), can be one of the most effective site preparations for biocrust recovery on unstable soils (Faist et al. 2020b; Zhou et al. 2020). Care is required when implementing soil stabilization practices because soil stabilizers and straw checkerboards can decrease biofilm establishment when soil is already stable (Antoninka et al. 2020a). Other effective methods of decreasing sand movement onto developing biocrusts include tree shelterbelts, shrubs, and sand fences (Zhou et al. 2020). Detailed manuals exist to guide biofilm restoration (Faist 2020a), and cyanobacterial crust restoration has been seen to be successful at the scale of hundreds of hectares (Zhou et al. 2020). Research is currently being conducted in the Salton Sea to determine the potential for biocrusts to stabilize emissive surfaces (Salton Sea Management Program 2024).

APPLICATION OF TRIBAL KNOWLEDGE FOR DUST CONTROL

If dust control measures are determined to be necessary and feasible on off-lake sources, Tribal input into the evaluation of potential control mechanisms, starting at the very initial stages of project conceptualization and design, could support collaborative planning, successful implementation, and community engagement. For example, in similar projects, collaboration with indigenous communities improved community investment in restoration efforts and outcomes (Gann et al. 2019). If parties liable for managing off-lake dust control measures engage with Tribal communities even before any federally mandated environmental review, they will learn from Tribal knowledge and build trust amongst the Tribal Nations in the OVPA.

As described previously, tribes historically engaged in extensive spreading of water across the valley floor north of the Owens Lake using irrigation ditches (Cheyenne Stone, Big Pine Paiute Tribe of the Owens Valley, personal communication, November 2024; Lawton et al. 1976). This practice could be a promising approach to promote vegetation cover and reduce dust emission. Additional Tribal input to the panel for mitigating PM_{10} emissions included mapping and clearing Russian thistle from ditches and roadways, as they can create dust once they become tumbleweeds. Limiting new excavations and new roadways east of Owens River and as well as limiting vehicle recreation on dry backwater lakes was also proposed as dust control methods (Mel Joseph, Lone Pine Paiute Shoshone Tribe, personal communication, August 2024).

CONCLUSIONS

If dust control measures are determined to be necessary and feasible for off-lake sources, implementation will require a systems-level landscape approach that considers cultural resources. Collaboration with local Tribal Nations will improve community investment in restoration efforts and outcomes.

Conclusion 6-1: Tribal input into the evaluation of potential dust control measures, starting at the very initial stages of project conceptualization and design, will support collaborative planning, community engagement, and successful implementation.

Many areas around the OVPA are extremely dynamic settings, requiring different approaches over space, and possible re-treatment over time (e.g., in flood deposits). Nevertheless, the panel found that the establishment and maintenance of vegetation offers the best chance for natural, self-sustaining protection of the soil surface from wind. Additional methods that will support the eventual establishment of vegetation are expanded upon below.

Conclusion 6-2: Establishing and maintaining native vegetation is the most stable and sustainable dust control measure across all emitting off-lake surfaces.

A number of sand sheets and dune fields (e.g., Keeler and Olancha) are distributed along the eastern and southern side of the shoreline. Efforts to partially stabilize Keeler Dunes using solid natural roughness elements of straw bales has resulted in the successful establishment of native shrub seedlings, which are providing seed for additional colonization of the sand. Porous naturally sourced roughness elements of straw checkerboards have also been used with great success in China to protect highways, rail lines, and villages from encroaching sand. These barriers are inexpensive to build and, although relatively short-lived, may be repaired or renewed as necessary until vegetation has successfully colonized the area.

Conclusion 6-3: In sand sheets and dune fields, solid naturally sourced roughness elements like straw bales and porous natural roughness elements like straw checkerboards are effective, ecologically favorable, and potentially feasible means to provide temporary surface stabilization until native shrub communities become well-established.

Fine-grained flood deposits are scattered in topographic lows along the shoreline and within the sand sheets and dunes. These fine-grained deposits are extremely emissive and efforts to control these dust sources could reduce exceedances. Where current deposits of fine-grained material are small in size, they could be covered with unlined layers of gravel or cobbles. This system would allow for rapid infiltration of water into the flood deposits that hold water very effectively. Thus, plants could find a hospitable root zone that would provide water and nutrients for growth and reproduction, further stabilizing the surface. Longevity of this dust control measure would depend on the interval between floods and would potentially need to be renewed following a major flood event barring changes to infrastructure or local topography.

Conclusion 6-4: For near-term mitigation of the highly emissive highway-impounded flood deposits, a feasible dust control measure is covering fine-grained flood deposits with gravel or cobbles in parallel with vegetation restoration.

Conclusion 6-5: The panel could not identify any long-term, cost-effective dust control measures that could stabilize the large-scale flood channel deposits deposited downgradient of the berm near Keeler Dunes by the remnants of Hurricane Kay in 2022.

Flood deposits around Owens Lake have been affected by both the construction of the highway and a number of upgradient berms. Drainage improvements along the highway with the addition of culverts or the elevation of roadways would reduce future accumulation of material behind highways leading to PM_{10} emissions. Highway berms concentrate flow toward a limited number of culverts under the highway, which may contribute to large deposits of fine-grained material after major precipitation events (e.g., the remnants of Hurricane Kay in 2022). Modification of the berm structures, highway culverts, and water harvesting and spreading are large, intensive dust management options, but they have the potential to reduce floodwater velocity, reduce the concentration of fine-grained sediment, and enhance water storage for vegetation establishment. Water harvesting from surface runoff and water spreading from drainage features could be used to encourage shrub growth in the future when climate change may make rainfall less predictable. Increased infiltration along the upper positions of the slope might also augment groundwater elevations such that the capillary fringe might be contacted by established shrub roots. Such a system may also limit erosion of gullies and ravines by reducing the velocity of floodwater.

Conclusion 6-6: Improved drainage for impounded flood deposits behind Highway 190 would reduce accumulation of fine-grained sediments from future flood events.

Conclusion 6-7: Hydrologic modifications of the berm structures, potentially combined with improved highway drainage and upgradient water harvesting and spreading, could reduce the size of future dust sources around Owens Lake.

Off-highway vehicle recreation is known to be associated with landscape changes that lead to PM_{10} exceedances. In the Owens Valley, the Bureau of Land Management manages an off-highway vehicle recreation and primitive camping area of approximately 400 acres (1.6 km^2) or 36 percent of the total area of the Olancha Dunes. Additionally, there have been reports of vehicle recreation on dry backwater lakes causing dust emissions.

Conclusion 6-8: Limits to recreational use, including off-highway vehicles, is a feasible dust control measure for recreational areas that contribute to PM_{10} exceedances.

TABLE 6-1 Reported Effectiveness, Performance, and Impacts of Dust Control Measures Potentially Applicable to Off-Lake Sources

Dust Control Measure	Reported Control Efficiency	Initial and Long-Term Water Use	Lifespan	Habitat Value	Impact to Cultural Resources	Time to Full Performance	Long term Performance	Site Suitability	Resilience to Extreme Meteorological Events and Climate Change
Sand Sheets and Dunes									
Artificial Roughness: Solid Natural	Depends on density and geometry. 92% observed at Keeler Dunes	With plants initially ~0.1 ft/yr. Without plants and after plant establishment, none. Lifespan unknown	Unknown	Low but may provide some cover for rodents	Potentially low land disturbance	Immediate	Will degrade with time if without plants. Long-term performance once plants are established	Suitable to sandy sites with significant saltation transport	Adaptable to all changes except flooding
Artificial Roughness: Porous Natural Plant Skeletons	Unknown, untested	None	Unknown, depends on availability and handling costs	Low	Potentially low land disturbance	Immediate	Will degrade with time	Suitable to small, culturally sensitive areas	Adaptable to all changes but flooding
Artificial Roughness: Porous Natural Straw Checkerboards	96–99 % reported	None	Approximately 3 years	Low	Incompatible: high land disturbance	Immediate	Degrades and may need replacement after 3 years, possibly longer in winter precipitation regime	Sand sheets and dunes	Adaptable to climate change but not to flooding
Artificial Roughness: Solid Engineered	Maximum of 90% ~200 ft from windward edge of array	None	Variable due to construction materials	Low	Potentially low land disturbance	Immediate	Depends on construction materials	Sand sheets and dunes	Adaptable to climate change but not flooding
Artificial Roughness: Porous Engineered	Unknown but increases with distance downwind into the array	None	Dependent on construction materials	Low	Potentially low land disturbance	Immediate	Depends on construction materials	Sand sheets and dunes	Adaptable to climate change but not flooding
Sand Control Fence	70–90%	None	20 yr lifespan	Low	Potentially low land disturbance	Immediate	Depends on construction materials	Sand sheets and dunes	Adaptable to climate change and flooding

ConMods	Total sand control unknown, but significantly greater native plant establishment	None	10 yr lifespan	Low	Compatible: low land disturbance	Years	Potentially very good	Sand sheets and sites needing more vegetation density	May be washed out in flash floods, otherwise unaffected by climate change
Perennial Shrub Communities	>95% with 20% shrub canopy cover (see Box 6-2)	Initially would require 0.1 ft/yr. Long-term would hopefully survive on natural precipitation	Permanent solution	Medium, provides cover and food	Incompatible: high land disturbance	Years to decades	Potentially very good	Sand sheets and flood deposits	Would be negatively and potentially severely impacted by long-term drought
Flood Deposits									
Covering with Gravel or Cobbles	99%	None	Long lifespan if not buried by blowing sand or flood deposited alluvium	Low	Incompatible: high land disturbance	Immediate	Gravel coverage can be renewed if buried by blowing sand or flood deposited alluvium	Suitable to fine sediments in flood deposited alluvium	Adaptable to climate change but not to flooding
Modified Highway-Associated Flood Management Infrastructure	Unknown	None	Long if designed to resist burial or clogging	Would facilitate shrub establishment on larger flood deposits	Not incompatible. Damage has already been done by previous highway construction	Immediate	Elevated roadways would probably be serviceable longer than distributed culverts	Flood prone areas made more emissive due to highway infrastructure	Design would need to consider extreme flows
Upslope Runoff Management	Variable based on slope and sediment composition	None	Decades	Enhances shrub and other plant establishment	Potentially high land disturbance locally	Immediate	Very good, especially after rills and gullies above have been filled with sediment	Areas upslope of recurrent flood deposits	Potentially impacted by extreme flows

125

References

Abatzoglou, John T., David S. Battisti, A. Park Williams, Winslow D. Hansen, Brian J. Harvey, and Crystal A. Kolden. 2021. "Projected Increases in Western US Forest Fire Despite Growing Fuel Constraints." *Communications Earth & Environment* 2 (1):1–8.

Alfano, Brigida, Luigi Barretta, Antonio Del Giudice, Saverio De Vito, Girolamo Di Francia, Elena Esposito, Fabrizio Formisano, Ettore Massera, Maria Lucia Miglietta, and Tiziana Polichetti. 2020. "A Review of Low-Cost Particulate Matter Sensors from the Developers' Perspectives." *Sensors* 20 (23):6819. https://doi.org/10.3390/s20236819.

American Geological Institute. 1983. *Dictionary of Geological Terms*. 3rd ed. New York: Anchor/Doubleday.

Antoninka, Anita, Matthew A. Bowker, Nichole N. Barger, Jayne Belnap, Ana Giraldo-Silva, Sasha C. Reed, Ferran Garcia-Pichel, and Michael C. Duniway. 2020a. "Addressing Barriers to Improve Biocrust Colonization and Establishment in Dryland Restoration." *Restoration Ecology* 28:S150–S159.

Antoninka, Anita, Akasha Faist, Emilio Rodriguez-Caballero, Kristina E. Young, V. Bala Chaudhary, Lea A. Condon, and David A. Pyke. 2020b. "Biological Soil Crusts in Ecological Restoration: Emerging Research and Perspectives." *Restoration Ecology* 28:S3–S8.

Armbrust, D. V. 1977. "A Review of Mulches to Control Wind Erosion." *Transactions of the ASAE* 20 (5):904–905.

Bacon, Steven N., Raymond M. Burke, Silvio K. Pezzopane, and Angela S. Jayko. 2006. "Last Glacial Maximum and Holocene Lake Levels of Owens Lake, Eastern California, USA." *Quaternary Science Reviews* 25 (11–12):1264–1282.

Bacon, Steven N., Nicholas Lancaster, Scott Stine, Edward J. Rhodes, and Grace A. McCarley Holder. 2018. "A Continuous 4000-Year Lake-Level Record of Owens Lake, South-Central Sierra Nevada, California, USA." *Quaternary Research* 90 (2):276–302.

Bacon, Steven N., Angela S. Jayko, Lewis A. Owen, Scott C. Lindvall, Edward J. Rhodes, Rina A. Schumer, and David L. Decker. 2020. "A 50,000-year Record of Lake-Level Variations and Overflow from Owens Lake, Eastern California, USA." *Quaternary Science Reviews* 238:106312. https://doi.org/10.1016/j.quascirev.2020.106312.

Baddock, Matthew C., Ted M. Zobeck, R. Scott Van Pelt, and Ed L. Fredrickson. 2011. "Dust Emissions from Undisturbed and Disturbed, Crusted Playa Surfaces: Cattle Trampling Effects." *Aeolian Research* 3 (1):31–41. https://doi.org/10.1016/j.aeolia.2011.03.007.

Baddock, Matthew C., Paul Ginoux, Joanna E. Bullard, and Thomas E. Gill. 2016. "Do MODIS-Defined Dust Sources Have a Geomorphological Signature?" *Geophysical Research Letters* 43 (6):2606–2613. https://doi.org/10.1002/2015GL067327.

Bagnold, Ralph A. 1941. "The Effect of Sand Movement on the Surface Wind." In *The Physics of Blown Sand and Desert Dunes*, 57–76. Springer.

Barone, J. B., B. H. Kusko, L. L. Ashbaugh, and T. A. Cahill. 1979. *A Study of Ambient Aerosols in the Owens Valley Area*. University of California, Davis: Air Quality Group Crocker Nuclear Laboratory.

REFERENCES

Barone, J. B., L. L. Ashbaugh, B. H. Kusko, and T. A. Cahill. 1981. "The Effect of Owens Dry Lake on Air Quality in the Owens Valley with Implications for the Mono Lake Area." In *Atmospheric Aerosol*, 327–346. American Chemical Society.

Basagic, Hassan Jules, and A. G. Fountain. 2011. "Quantifying 20th Century Glacier Change in the Sierra Nevada, California." *Arctic, Antarctic, and Alpine Research* 43 (3):317–330.

Batterman, Stuart, Rajiv Ganguly, Vlad Isakov, Janet Burke, Saravanan Arunachalam, Michelle Snyder, Thomas Robins, and Toby Lewis. 2014. "Dispersion Modeling of Traffic-Related Air Pollutant Exposures and Health Effects Among Children with Asthma in Detroit, Michigan." *Transportation Research Record* 2452 (1):105–113.

Bellasio, Roberto, Roberto Bianconi, Sonia Mosca, and Paolo Zannetti. 2017. "Formulation of the Lagrangian Particle Model LAPMOD and Its Evaluation against Kincaid SF6 and SO2 Datasets." *Atmospheric Environment* 163:87–98.

Belnap, Jayne. 2006. "The Potential Roles of Biological Soil Crusts in Dryland Hydrologic Cycles." *Hydrological Processes: An International Journal* 20 (15):3159–3178.

Belnap, Jayne, Susan L. Phillips, J. E. Herrick, and J. R. Johansen. 2007. "Wind Erodibility of Soils at Fort Irwin, California (Mojave Desert), USA, Before and After Trampling Disturbance: Implications for Land Management." *Earth Surface Processes and Landforms: The Journal of the British Geomorphological Research Group* 32 (1):75–84.

Benson, Larry, Michaele Kashgarian, Robert Rye, Steve Lund, Fred Paillet, Joseph Smoot, Cynthia Kester, Scott Mensing, Dave Meko, and Susan Lindström. 2002. "Holocene Multidecadal and Multicentennial Droughts Affecting Northern California and Nevada." *Quaternary Science Reviews* 21 (4-6):659–682.

Bestelmeyer, Brandon T., Gregory S. Okin, Michael C. Duniway, Steven R. Archer, Nathan F. Sayre, Jebediah C. Williamson, and Jeffrey E. Herrick. 2015. "Desertification, Land Use, and the Transformation of Global Drylands." *Frontiers in Ecology and the Environment* 13 (1):28–36.

Bilbro, J. D., and J. E. Stout. 1999. "Wind Velocity Patterns as Modified By Plastic Pipe Windbarriers." *Journal of Soil and Water Conservation* 54 (3):551–556.

Birmili, W., K. Schepanski, A. Ansmann, G. Spindler, I. Tegen, B. Wehner, A. Nowak, E. Reimer, I. Mattis, K. Müller, E. Brüggemann, T. Gnauk, H. Herrmann, A. Wiedensohler, D. Althausen, A. Schladitz, T. Tuch, and G. Löschau. 2008. "A Case of Extreme Particulate Matter Concentrations Over Central Europe Caused by Dust Emitted Over The Southern Ukraine." *Atmospheric Chemistry and Physics* 8 (4):997–1016. https://doi.org/10.5194/acp-8-997-2008.

Blanton, Stephen G., Katheryn R. Kolesar, and David A. Jaffe. 2022. "Contributions of Sediment From the Slate Canyon Alluvial Fan to the Formation and Morphogenesis of the Keeler Dunes, CA." *Frontiers in Earth Science* 10:879115.

Bo, Tian-Li, Peng Ma, and Xiao-Jing Zheng. 2015. "Numerical Study on the Effect of Semi-Buried Straw Checkerboard Sand Barriers on the Wind Speed." *Aeolian Research* 16:101–107. http://dx.doi.org/10.1016/j.aeolia.2014.10.002.

Boers, Th. M., and Jiftah Ben-Asher. 1982. "A Review of Rainwater Harvesting." *Agricultural Water Management* 5 (2):145–158.

Borgias, Sophia L. 2020. *Public interest, Indigenous rights, and the Los Angeles Aqueduct*. PhD diss., University of Arizona.

Borgias, Sophia L. 2024. "Drought, Settler Law, and the Los Angeles Aqueduct: The Shifting Political Ecology of Water Scarcity in California's Eastern Sierra Nevada." *Journal of Political Ecology* 31 (1).

Bowker, Matthew A, Jayne Belnap, Burkhard Büdel, Christophe Sannier, Nicole Pietrasiak, David J. Eldridge, and Víctor Rivera-Aguilar. 2016. "Controls on Distribution Patterns of Biological Soil Crusts at Micro-To Global Scales." *Biological Soil Crusts: An Organizing Principle in Drylands*:173–197.

Bullard, Joanna, and Matthew Baddock. 2019. "Dust: Sources, Entrainment, Transport." In *Aeolian Geomorphology: A New Introduction*, 81–106.

Carmichael, Gregory R., Adrian Sandu, Tianfeng Chai, Dacian N. Daescu, Emil M. Constantinescu, and Youhua Tang. 2008. "Predicting Air Quality: Improvements through Advanced Methods to Integrate Models and Measurements." *Journal of Computational Physics* 227 (7):3540–3571.

Cahill, T. A., T. E. Gill, D. A. Gillette, E. A. Gearhart, J. S. Reid, and M.L. Yau. 1994. *Generation, characterization, and transport of Owens (Dry) Lake dusts*. Final Report, Contract A132-105. Sacramento: California Air Resources Board.

CARB (California Air Resources Board). 2024. California Ambient Air Quality Standards. https://ww2.arb.ca.gov/resources/california-ambient-air-quality-standards.

California Department of Water Resources Climate Change Technical Advisory Group. 2015. "Perspectives and Guidance for Climate Change Analysis." *California Department of Water Resources Technical Information Record*:142.

California Natural Resource Agency. 2024. *Annual Report on the Salton Sea Management Program*. https://saltonsea.ca.gov/wp-content/uploads/2021/03/2021-Annual-Report_3-5-21.pdf

Chauhan, Amit Kumar Singh, Anant Singh Suryavanshi, Anoushka Awasthi, Shivani Rai, Shruti Mishra, and Arun Kumar Singh. 2022. "Solar Powered Low Cost Air Quality Monitoring System." International Conference on VLSI, Communication and Signal Processing.

Chen, Hao, Song Bai, Douglas Eisinger, Deb Niemeier, and Michael Claggett. 2009. "Predicting Near-Road PM2. 5 Concentrations: Comparative Assessment of CALINE4, CAL3QHC, and AERMOD." *Transportation Research Record* 2123 (1):26–37.

Chen, Yilin, Huizhong Shen, Jennifer Kaiser, Yongtao Hu, Shannon L. Capps, Shunliu Zhao, Amir Hakami, Jhih-Shyang Shih, Gertrude K. Pavur, and Matthew D. Turner. 2021. "High-Resolution Hybrid Inversion of IASI Ammonia Columns to Constrain US Ammonia Emissions Using the CMAQ Adjoint Model." *Atmospheric Chemistry and Physics* 21 (3):2067–2082.

Cheung, Suet-Yi, Ian J. Walker, Soe W. Myint, and Ronald I. Dorn. 2021. "Assessing Land Degradation Induced by Recreational Activities in the Algodones Dunes, California Using MODIS Satellite Imagery." *Journal of Arid Environments* 185:104334. https://doi.org/10.1016/j.jaridenv.2020.104334.

Chimner, Rodney A, and David J. Cooper. 2004. "Using Stable Oxygen Isotopes to Quantify the Water Source used for Transpiration by Native Shrubs in the San Luis Valley, Colorado USA." *Plant and Soil* 260:225–236.

Ciren, Pubu, Shobha Kondragunta, and Amy Huff. 2024. "ABI Enterprise Processing System Aerosol Detection Product: Algorithm Theoretical Basis Document." *Applications and Research* 1.1.

Clifford, Katherine R. 2022. "Natural Exceptions or Exceptional Natures? Regulatory Science and the Production of Rarity." *Annals of the American Association of Geographers* 112 (8):2287–2304.

Colledge, Michelle A., Jaime R. Julian, Vihra V. Gocheva, Cheryl L. Beseler, Harry A. Roels, Danelle T. Lobdell, and Rosemarie M. Bowler. 2015. "Characterization of Air Manganese Exposure Estimates for Residents in Two Ohio Towns." *Journal of the Air & Waste Management Association* 65 (8):948–957.

Cooper, David J., John S. Sanderson, David I. Stannard, and David P. Groeneveld. 2006. "Effects of Long-Term Water Table Drawdown on Evapotranspiration and Vegetation in an Arid Region Phreatophyte Community." *Journal of Hydrology* 325 (1):21–34. https://doi.org/10.1016/j.jhydrol.2005.09.035.

Coulter, Thomas C. 2004. *EPA-CMB 8.2 User's Manual*. Research Triangle Park, NC: Air Quality Modeling Group.

Cowan, Jill. 2024. "In California, Tribal Members Are Reclaiming the 'Land of the Flowing Water.'" *New York Times*. https://www.nytimes.com/2024/06/16/us/california-native-american-tribes.html?searchResultPosition=1.

Dagliya, Monika, and Neelima Satyam. 2024. "Large Scale Study on Influence of Biopolymer to Mitigate Wind Induced Sand Erosion with Durability Analysis." *Soil and Tillage Research* 236:105942. https://doi.org/10.1016/j.still.2023.105942.

Danskin, Wesley R. 1998. *Evaluation of the Hydrologic System and Selected Water-Management Alternatives in the Owens Valley, California*. In Water Supply Paper 2370-H. Reston, VA: U.S. Geological Survey.

David, Liji M., A. R. Ravishankara, Steven J. Brey, Emily V. Fischer, John Volckens, and Sonia Kreidenweis. 2021. "Could the Exception Become the Rule? 'Uncontrollable' Air Pollution Events in the US Due to Wildland Fires." *Environmental Research Letters* 16 (3):034029.

Dehkordi, Elahe Ahmadpoor, Ali Abbasi Surki, Mehdi Pajouhesh, and Pejman Tahmasebi. 2022. "Straw checkerboard barriers improve soil restoration and mitigate the impacts of drought on Medicago scutellata L." *Ecological Engineering* 178: 106578. https://doi.org/10.1016/j.ecoleng.2022.106578. https://ui.adsabs.harvard.edu/abs/2022EcEng.17806578A.

Dehkordi, Elahe Ahmadpoor, Ali Abbasi Surki, Mehdi Pajouhesh, and Pejman Tahmasebi. 2023. "Ecological Restoration of Sloping Land Using Straw Checkerboard Barriers Seeded With Winter Cover Crops." *Ecological Engineering* 193. https://doi.org/10.1016/j.ecoleng.2023.106994.

Dettinger, Michael D., Holly Alpert, John J. Battles, Jonathan Kusel, Hugh Safford, Dorian Fougeres, Clarke Knight, Lauren Miller, and Sarah Sawyer. 2018. "Sierra Nevada Summary Report. California's Fourth Climate Change Assessment." https://www.energy.ca.gov/sites/default/files/2019-11/Reg_Report-SUM-CCCA4-2018-004_SierraNevada_ADA.pdf

Devitt, Dale A., and Brian Bird. 2016. "Changes in groundwater oscillations, soil water content and evapotranspiration as the water table declined in an area with deep rooted phreatophytes." *Ecohydrology* 9 (6):1082–1093.

Dong, Zhibao, Donald W. Fryrear, and Shangyu Gao. 1999. "Modeling the Roughness Properties of Artificial Soil Clods." *Soil Science* 164 (12):930–935.

Donnallan, Andrea, Bernard Hallet, and Sebastien Leprince. 2015. *Gazing at the Solar System: Capturing the Evolution of Dunes, Faults, Volcanoes, and Ice from Space*. California Institute of Technology.

Donovan, Lisa A., James H. Richards, and Matthew W. Muller. 1996. "Water Relations and Leaf Chemistry of *Chrysothamnus Nauseosus* ssp. *Consimilis* (Asteraceae) and *Sarcobatus vermiculatus* (Chenopodiaceae)." *American Journal of Botany* 83 (12):1637–1646.

East, Amy E., and Joel B. Sankey. 2020. "Geomorphic and Sedimentary Effects of Modern Climate Change: Current and Anticipated Future Conditions in the Western United States." *Reviews of Geophysics* 58 (4). https://doi.org/10.1029/2019RG000692.

Eaton, Gordon P. 1982. "The Basin and Range Province: Origin and Tectonic Significance." *Annual Review of Earth and Planetary Sciences* 10:409.

REFERENCES

Eibedingil, Iyasu G., Thomas E. Gill, R. Scott Van Pelt, and Daniel Q. Tong. 2021. "Comparison of Aerosol Optical Depth from MODIS Product Collection 6.1 and AERONET in the Western United States." *Remote Sensing* 13 (12):2316.

Eibedingil, Iyasu G., Thomas E. Gill, Tarek Kandakji, Jeffrey A. Lee, Junran Li, and R. Scott Van Pelt. 2024. "Effect of Spatial and Temporal 'Drought Legacy' on Dust Sources in Adjacent Ecoregions." *Land Degradation & Development* 35 (4):1511–1525.

Elmore, Andrew J., John F. Mustard, and Sara J. Manning. 2003. "Regional Patterns of Plant Community Response to Changes in Water: Owens Valley, California." *Ecological Applications* 13 (2):443-460.

Elmore, Andrew J., James M. Kaste, Gregory S. Okin, and Matthew S. Fantle. 2008. "Groundwater Influences on Atmospheric Dust Generation in Deserts." *Journal of Arid Environments* 72 (10):1753–1765. https://doi.org/10.1016/j.jaridenv.2008.05.008.

Elsner, James B., and Anastasios A. Tsonis. 1996. *Singular Spectrum Analysis: A New Tool in Time Series Analysis*. New York: Springer Science + Business Media.

Environmental Science Associates. 2022. *Salton Sea Monitoring Implementation Plan (SSMIP)*. https://saltonsea.ca.gov/wp-content/uploads/2022/02/00_SaltonSeaMIP_508-Final.pdf.

EPA (U.S. Environmental Protection Agency). 2017. "Owens Valley Particulate Matter Plan Q & A." https://19january2017snapshot.epa.gov/www3/region9/air/owens/qa.html.

EPA. 2019a. *Exceptional Events Guidance: Prescribed Fire on Wildland that May Influence Ozone and Particulate Matter Concentrations*. https://www.epa.gov/sites/default/files/2019-08/documents/ee_prescribed_fire_final_guidance_-_august_2019.pdf.

EPA. 2019b. "Guidance on the Preparation of Demonstrations in Support of Requests to Exclude Ambient Air Quality Data Influenced by High Wind Dust Events Under the 2016 Exceptional Events Rule." https://www.epa.gov/sites/default/files/2019-04/documents/high_wind_dust_event_guidance.pdf.

EPA. 2019c. *Meteorological Model Performance for Annual 2016 Simulation WRF v3.8*. Research Triangle Park, NC. https://www.epa.gov/sites/default/files/2020-10/documents/met_model_performance-2016_wrf.pdf.

EPA. 2024a. "Air Quality Dispersion Modeling." https://www.epa.gov/scram/air-quality-dispersion-modeling.

EPA. 2024b. "Air Quality Dispersion Modeling—Preferred and Recommended Models." https://www.epa.gov/scram/air-quality-dispersion-modeling-preferred-and-recommended-models.

EPA. 2025a. "Air Quality Design Values." https://www.epa.gov/air-trends/air-quality-design-values.

EPA. 2025b. "CMAQ: The Community Multiscale Air Quality Modeling System." https://www.epa.gov/cmaq.

Espinoza, Vicky, Duane E. Waliser, Bin Guan, David A. Lavers, and F. Martin Ralph. 2018. "Global Analysis of Climate Change Projection Effects on Atmospheric Rivers." *Geophysical Research Letters* 45 (9):4299–4308. https://doi.org/10.1029/2017GL076968.

Etyemezian, Vicken, G. Nikolich, S. Ahonen, Marc Pitchford, Mark Sweeney, R. Purcell, J. Gillies, and Hampden Kuhns. 2007. "The Portable In Situ Wind Erosion Laboratory (PI-SWERL): A new method to measure PM_{10} windblown dust properties and potential for emissions." *Atmospheric Environment* 41 (18):3789–3796. https://doi.org/10.1016/j.atmosenv.2007.01.018.

Faist, Akasha, Colin Tucker, Sasha Reed, Anita Antoninka, Matthew Bowker, Nichole Barger, Kara Dohrenwend, Natalie Day, Sue Bellagamba, Jayne Belnap, Michael Duniway, Stephen Fick, Ana Giraldo-Silva, Corey Nelson, Julie Bethany, Sergio Velasco-Ayuso, Ferran Garcia-Pichel. 2020a. "Operational Manual for Biocrust Restoration in Drylands." https://anitaantoninka.wixsite.com/biocrustrestoration.

Faist, Akasha, Anita J. Antoninka, Jayne Belnap, Matthew A. Bowker, Michael C. Duniway, Ferran Garcia-Pichel, Corey Nelson, Sasha C. Reed, Ana Giraldo-Silva, and Sergio Velasco-Ayuso. 2020b. "Inoculation and Habitat Amelioration Efforts in Biological Soil Crust Recovery Vary by Desert and Soil Texture." *Restoration Ecology* 28:S96–S105.

FAO (Food and Agriculture Organization). 1991. *Water Harvesting: A Manual for the Design and Construction of Water Harvesting Schemes For Plant Production*. W. Critchley and K. Siegert, editors. *Rome: UN Food and Agriculture Organization*.

Fick, Stephen E., Natalie Day, Michael C. Duniway, Sean Hoy-Skubik, and Nichole N. Barger. 2020. "Microsite Enhancements for Soil Stabilization and Rapid Biocrust Colonization in Degraded Drylands." *Restoration Ecology* 28:S139–S149.

Field, Jason P., Jayne Belnap, David D. Breshears, Jason C. Neff, Gregory S. Okin, Jeffrey J. Whicker, Thomas H. Painter, Sujith Ravi, Marith C. Reheis, and Richard L. Reynolds. 2010. "The Ecology of Dust." *Frontiers in Ecology and the Environment* 8 (8):423–430.

Finnigan, John J. 1988. "Air Flow Over Complex Terrain." In *Flow and Transport in the Natural Environment: Advances and Applications*, edited by William L. Steffen and Owen T. Denmead, 183–229. Heidelberg, Germany: Springer.

Frie, Alexander L., Justin H. Dingle, Samantha C. Ying, and Roya Bahreini. 2017. "The Effect of a Receding Saline Lake (the Salton Sea) on Airborne Particulate Matter Composition." *Environmental Science & Technology* 51 (15):8283–8292.

Frie, Alexander L., Alexis C. Garrison, Michael V. Schaefer, Steve M. Bates, Jon Botthoff, Mia Maltz, Samantha C. Ying, Timothy Lyons, Michael F. Allen, and Emma Aronson. 2019. "Dust sources in the Salton Sea Basin: A clear case of an anthropogenically impacted dust budget." *Environmental Science & Technology* 53 (16): 9378-9388.

Friedlander, Sheldon K. 1973. "Chemical Element Balances and Identification of Air Pollution Sources." *Environmental Science & Technology* 7 (3):235–240.

Gale, Hoyt Stoddard. 1914. *Salines in the Owens, Searles, and Panamint Basins, Southeastern California*. US Government Printing Office.

Gann, George D., Tein McDonald, Bethanie Walder, James Aronson, Cara R. Nelson, Justin Jonson, James G. Hallett, Cristina Eisenberg, Manuel R. Guariguata, and Junguo Liu. 2019. "International Principles and Standards for the Practice of Ecological Restoration." *Restoration Ecology* 27 (S1):S1–S46.

Ganskopp, David C. 1986. "Tolerances of sagebrush, rabbitbrush, and greasewood to elevated water tables." *Journal of Range Management* 39 (4):334–337. doi: 10.2307/3899774.

Garcia, C. Amanda, Jena M. Huntington, Susan G. Buto, Michael T. Moreo, J. LaRue Smith, and Brian J. Andraski. 2015. *Groundwater Discharge by Evapotranspiration, Dixie Valley, West-Central Nevada, March 2009–September 2011*. Reston, VA: US Geological Survey.

GBUAPCD (Great Basin Unified Air Pollution Control District). 1998. "Owens Valley PM_{10} Planning Area Demonstration of Attainment, State Implementation Plan." https://ww2.arb.ca.gov/resources/documents/1998-owens-valley-pm10-attainment-demonstration-sip.

GBUAPCD. 2003. "Owens Valley PM_{10} Planning Area Demonstration of Attainment, State Implementation Plan. 2003 Revision." https://ww2.arb.ca.gov/resources/documents/2003-owens-valley-pm10-sip-revision.

GBUAPCD. 2008. "Owens Valley PM_{10} Planning Area Demonstration of Attainment, State Implementation Plan." https://ww2.arb.ca.gov/resources/documents/2008-owens-valley-pm10-attainment-demonstration-sip.

GBUAPCD. 2013. "2013 Amendment to the Owens Valley PM_{10} SIP (Board Order #130916-01)." https://ww2.arb.ca.gov/resources/documents/2013-owens-valley-pm10-sip-revision.

GBUAPCD. 2016. "2016 Owens Valley Planning Area PM_{10} State Implementation Plan." https://ww2.arb.ca.gov/resources/documents/2016-owens-valley-pm10-sip-revision.

GBUAPCD. 2023. "Hearing Board Meeting Information for the Novermber 3, 2023 Meeting." https://www.gbuapcd.org/Docs/District/PublicNotice/2023/20231030_HearingBoardPacket_InterimVarianceGB23-01.pdf.

GBUAPCD. 2024a. "Air Quality Camera Map.".https://www.gbuapcd.org/AirMonitoringData/AirQualityCameras/Map.

GBUAPCD. 2024b. "Annual Air Quality Monitoring Network Plan Draft." https://www.gbuapcd.org/Docs/District/PublicNotice/2024/20240607_2024AMNP.pdf.

Gershunov, Alexander, Tamara Shulgina, Rachel E. S. Clemesha, Kristen Guirguis, David W. Pierce, Michael D. Dettinger, David A. Lavers, Daniel R. Cayan, Suraj D. Polade, and Julie Kalansky. 2019. "Precipitation Regime Change in Western North America: The Role of Atmospheric Rivers." *Scientific Reports* 9 (1):9944.

Gifford, Franklin A., and Steven R. Hanna. 1970. *Urban air pollution modelling*. Oak Ridge, TN: U.S. Department of Commerce and the National Oceanic and Atmospheric Administration.

Gifford, Franklin A., and Steven R. Hanna. 1973. "Modelling Urban Air Pollution." *Atmospheric Environment* (1967) 7 (1):131–136.

Gill, Thomas Edward. 1995. "Dust Generation Resulting from Desiccation of Playa Systems: Studies on Mono and Owens Lakes, California." PhD diss, University of California, Davis.

Gillette, Dale A., John Adams, Albert Endo, Dudley Smith, and Rolf Kihl. 1980. "Threshold Velocities for Input of Soil Particles into the Air By Desert Soils." *Journal of Geophysical Research: Oceans* 85 (C10):5621–5630.

Gillette, Dale A., D. W. Fryrear, Thomas E. Gill, Trevor Ley, Thomas A. Cahill, and Elizabeth A. Gearhart. 1997. "Relation of Vertical Flux of Particles Smaller than 10 μm to Total Aeolian Horizontal Mass Flux at Owens Lake." *Journal of Geophysical Research: Atmospheres* 102 (D22):26009–26015. https://doi.org/10.1029/97JD02252.

Gillette, Dale A., Duane Ono, and Kenneth Richmond. 2004. "A Combined Modeling and Measurement Technique for Estimating Windblown Dust Emissions at Owens (Dry) Lake, California." *Journal of Geophysical Research: Earth Surface* 109 (F1). https://doi.org/10.1029/2003JF000025.

Gillies, John A., and Nicholas Lancaster. 2013. "Large Roughness Element Effects on Sand Transport, Oceano Dunes, California." *Earth Surface Processes and Landforms* 38 (8):785–792.

Gillies, John A., H. Green, G. McCarley-Holder, S. Grimm, C. Howard, N. Barbieri, D. Ono, and T. Schade. 2015. "Using Solid Element Roughness to Control Sand Movement: Keeler Dunes, Keeler, California." *Aeolian Research* 18:35–46.

Gillies, John A., Vicken Etyemezian, George Nikolich, and William G. Nickling. 2017. *The Engineered Roughness Experiment: Owens Lake, CA*. Bishop, CA: Great Basin Unified Air Pollution Control District.

REFERENCES

Gillies, John A., Vicken Etyemezian, and George Nikolich. 2018a. "Trapping of Sand-Sized Particles Exterior and Interior to Large Porous Roughness Forms in the Atmospheric Surface Layer: Phases 1 & 2." *Boundary-Layer Meteorology* 170(3).

Gillies, John A., Vicken Etyemezian, George Nickolich, William G. Nickling, and Jasper F. Kok. 2018b. "Changes in the Saltation Flux Following Step-Change in Macro-Roughness." *Earth Surface Processes & Landforms* 43 (9):1871–1844. http://doi.org/10.1002/esp.4362.

Gillies, John A., E. Furtak-Cole, and Vicken Etyemezian. 2020. "Increments of Progress Towards Air Quality Objectives-ODSVRA Dust Controls." Desert Research Institute. https://ohv.parks.ca.gov/pages/1140/files/08-26-2021-Item%20 4C-Oceano%20Dunes%20SVRA%20Increments%20of%20Progress%20(Attachment).pdf.

Gillies, John A., E. Furtak-Cole, George Nikolich, and Vicken Etyemezian. 2022. "The Role of Off-Highway Vehicle Activity in Augmenting Dust Emissions at the Oceano Dunes State Vehicular Recreation Area, Oceano, CA." *Atmospheric Environment: X* 13:100146. https://doi.org/10.1016/j.aeaoa.2021.100146.

Goedhart, C. M., and D. E. Pataki. 2011. "Ecosystem Effects of Groundwater Depth in Owens Valley, California." *Ecohydrology* 4 (3):458–468.

Goudie, Andrew. 2011. "Parabolic Dunes: Distribution, Form, Morphology and Change." *Annals of Arid Zone* 50 (3&4):1–7.

Graff, Arno, David Strimaitis, and Robert Yamartino. 1998. "Regulatory-Oriented Features of the Kinematic Simulation Particle Model." In *Air Pollution Modeling and Its Application XII*, 655-664. Springer.

Grell, Georg A., Steven E. Peckham, Rainer Schmitz, Stuart A. McKeen, Gregory Frost, William C. Skamarock, and Brian Eder. 2005. "Fully Coupled "Online" Chemistry within the WRF model." *Atmospheric Environment* 39 (37):6957–6975.

Griepentrog, T. E., and D. P. Groeneveld. 1981. *The Owens Valley Management Report: Final Report for Inyo County, Bishop, CA*.

Groom, Jeremiah D., Lloyd B. McKinney, Lianne C. Ball, and Clark S. Winchell. 2007. "Quantifying Off-Highway Vehicle Impacts on Density and Survival of a Threatened Dune-Endemic Plant." *Biological Conservation* 135 (1):119–134. https://doi.org/10.1016/j.biocon.2006.10.005.

Hagan, David. 2022. "Can your Plantower PMS5003-Based Air Quality Sensor Measure PM_{10}?" QuantAQ. https://blog.quant-aq.com/can-your-plantower-pms5003-based-air-quality-sensor-measure-pm10.

Hagan, David, and Jesse H. Kroll. 2020. "Assessing the Accuracy of Low-Cost Optical Particle Sensors Using a Physics-Based Approach." *Atmospheric Measurement Techniques* 13 (11):6343–6355. https://doi.org/10.5194/amt-2020-188.

Hakami, A., D. K. Henze, J. H. Seinfeld, T. Chai, Y. Tang, G. R. Carmichael, and A. Sandu. 2005. "Adjoint Inverse Modeling of Black Carbon during the Asian Pacific Regional Aerosol Characterization Experiment." *Journal of Geophysical Research: Atmospheres* 110 (D14301).

Halopka, Richard. 2022. "Hay Market Demand and Price Report for the Upper Midwest–for February 14, 2022." *University of Wisconsin*.

Hanna, Steven R. 1971. "A Simple Method of Calculating Dispersion from Urban Area Sources." *Journal of the Air Pollution Control Association* 21 (12):774–777.

Harpold, Adrian A., Noah P. Molotch, Keith N. Musselman, Roger C. Bales, Peter B. Kirchner, Marcy Litvak, and Paul D. Brooks. 2015. "Soil Moisture Response to Snowmelt Timing in Mixed-Conifer Subalpine Forests." *Hydrological Processes* 29 (12):2782–2798. https://doi.org/10.1002/hyp.10400.

Haverstock, Gregory J., Angela S. Jayko, and Harry C. Williams. 2022. "The Archaeological Identification and Radiocarbon Assay of Pre-Colonial Nüümü (Paiute) Agriculture in Payahuunadü Owens Valley), California." *Journal of California and Great Basin Anthropology* 42:23–40.

Hesp, Patrick A., and Thomas A. G. Smyth. 2019. "Anchored Dunes." In *Aeolian Geomorphology: A New Introduction*, 157–178.

Holder, Grace. 1997. "Off-Lake Dust Sources, Owens Lake Basin." Great Basin Unified Air Pollution Control District. Bishop, CA.

Holder, Grace, Sondra Grimm, Ann Logan, Steven Bacon, and Scott Warner. 2024. "Comment on the Formation of the Keeler Dunes: Aeolian Research articles, Vol 54, 100764, 100765, and 100773 and Aeolian Research article, Vol 58 100819." Great Basin Unified Air Pollution Control District. Bishop, CA.

Hollett, Kenneth J., Wesley R. Danskin, William F. McCaffrey, and Caryl L. Walti. 1989. *Geology and water resources of Owens Valley, California*. Open-File Report 88-715. Reston, VA: US Geological Survey.

Hollett, Kenneth J., Wesley R. Danskin, William F. McCaffrey, and Caryl L. Walti. 1991. *Geology and Water Resources of Owens Valley, California*. Water-Supply Paper 2370-B. Reston, VA: US Geological Survey.

House, P. Kyle, Brenda J. Buck, and Alan R. Ramelli. 2010. *Geologic Assessment of Piedmont and Playa Flood Hazards in the Ivanpah Valley Area, Clark County, Nevada*. Nevada Bureau of Mines and Geology Report 53.

Huang, Xingying, Samantha Stevenson, and Alex D. Hall. 2020. "Future Warming and Intensification of Precipitation Extremes: A "Double Whammy" Leading to Increasing Flood Risk in California." *Geophysical Research Letters* 47 (16):e2020GL088679.

Jabis, Meredith D. 2011. "Owens Valley Vegetation Conditions 2010." Inyo County Water Department Staff Report. https://www.inyowater.org/wp-content/uploads/legacy/Annual_Reports/2010_2011/documents/VEGETATIONFORWEB.pdf.

James, David E., Johan Pulgarin, Jon Becker, Sherrie Edwards, Tina Gingras, Gina Venglass, and Carrie MacDougall. 1999. *Development of Vacant Land Emission PM-10 Factors in the Las Vegas Valley*. https://www3.epa.gov/ttnchie1/conference/ei10/fugdust/james.pdf

Jiang, Qingfang, Ming Liu, and James D. Doyle. 2011. "Influence of Mesoscale Dynamics and Turbulence on Fine Dust Transport in Owens Valley." *Journal of Applied Meteorology and Climatology* 50 (1):20–38.

Jittra, Nattawut, Nattaporn Pinthong, and Sarawut Thepanondh. 2015. "Performance Evaluation of AERMOD and CALPUFF Air Dispersion Models in Industrial Complex Area." *Air, Soil and Water Research* 8.

Kaiser, Jennifer, Daniel J. Jacob, Lei Zhu, Katherine R. Travis, Jenny A. Fisher, Gonzalo González Abad, Lin Zhang, Xuesong Zhang, Alan Fried, and John D. Crounse. 2018. "High-Resolution Inversion of OMI Formaldehyde Columns to Quantify Isoprene Emission on Ecosystem-Relevant Scales: Application to the Southeast US." *Atmospheric Chemistry and Physics* 18 (8):5483–5497.

Kandakji, Tarek, Thomas E. Gill, and Jeffrey A. Lee. 2020. "Identifying and Characterizing Dust Point Sources in the Southwestern United States Using Remote Sensing and GIS." *Geomorphology* 353:107019.

Kandakji, Tarek, Thomas E. Gill, and Jeffrey A. Lee. 2021. "Drought and Land Use/Land Cover Impact on Dust Sources in Southern Great Plains and Chihuahuan Desert of the US: Inferring Anthropogenic Effect." *Science of the Total Environment* 755:142461.

Kaur, Kamaljeet, and Kerry E. Kelly. 2023. "Performance Evaluation of the Alphasense OPC-N3 and Plantower PMS5003 Sensor in Measuring Dust Events in the Salt Lake Valley, Utah." *Atmospheric Measurement Techniques* 16 (10):2455–2470. https://doi.org/10.5194/amt-16-2455-2023.

Khatei, Ganesh, Tobia Rinaldo, R. Scott Van Pelt, Paolo D'Odorico, and Sujith Ravi. 2024. "Wind Erodibility and Particulate Matter Emissions of Salt-Affected Soils: The Case of Dry Soils in a Low Humidity Atmosphere." *Journal of Geophysical Research: Atmospheres* 129 (1):e2023JD039576. https://doi.org/10.1029/2023JD039576.

Kim, Bok Chul, and Donald R. Lowe. 2004. "Depositional Processes of the Gravelly Debris Flow Deposits, South Dolomite Alluvial Fan, Owens Valley, California." *Geosciences Journal* 8:153–170.

Kocurek, Gary, and Jamie Nielson. 1986. "Conditions Favourable for the Formation of Warm-Climate Aeolian Sand Sheets." *Sedimentology* 33 (6):795–816.

Kolesar, Katheryn R., Matthew. Mavko, E. Burgess, N. Nguyen, and Mark D. Schaaf. 2022a. "A Modified Resultant Drift Potential for More Accurate Prediction of Sand Transportation in the Vicinity of the Keeler Dunes, California." *Aeolian Research* 58:100819.

Kolesar, Katheryn, Mark Schaaf, John W. Bannister, Maarten D. Schreuder, and Mica H. Heilmann. 2022b. "Characterization of Potential Fugitive Dust Emissions within the Keeler Dunes, an Inland Dune Field in the Owens Valley, California, United States." *Aeolian Research* 54:100765. https://doi.org/10.1016/j.aeolia.2021.100765.

Kuula, Joel, Timo Mäkelä, Minna Aurela, Kimmo Teinilä, Samu Varjonen, Óscar González, and Hilkka Timonen. 2020. "Laboratory Evaluation of Particle-Size Selectivity of Optical Low-Cost Particulate Matter Sensors." *Atmospheric Measurement Techniques* 13 (5):2413–2423. https://doi.org/10.5194/amt-13-2413-2020.

LADWP (Los Angeles Department of Water and Power). 2010. Owens Lake Habitat Management Plan. Los Angeles, CA: Los Angeles Department of Water and Power. Available at https://inyomonowater.org/wp-content/uploads/2011/09/HabitatMgmtPlan_OwensDryLake_LADWP.pdf, accessed May 14, 2025.

LADWP. 2012. *Final Report on the Owens Lake Groundwater Evaluation Project*. https://www.ladwp.com/sites/default/files/documents/Final_Report_on_the_OLGEP.pdf.

LADWP. 2023. *2023 Annual Owens Valley Report*. https://www.ladwp.com/sites/default/files/2023-09/2023%20FINAL%20OWENS%20VALLEY%20REPORT.pdf.

LADWP. 2024a. *Annual Owens Valley Report*. https://www.ladwp.com/sites/default/files/2024-06/2024%20Final%20Owens%20Valley%20Report.pdf.

LADWP. 2024b. "Owens Lake." https://www.ladwp.com/who-we-are/water-system/los-angeles-aqueduct/owens-lake.

LADWP and County of Inyo. 1990a. "Agreement Between the County of Inyo and the City of Los Angeles and Its Department of Water and Power on a Long Term Groundwater Management Plan for Owens Valley and Inyo County." https://www.inyowater.org/documents/governing-documents/water-agreement.

REFERENCES

LADWP and County of Inyo. 1990b. "Water from the Owens Valley to Supply the Second Los Angeles Aqueduct." *Draft Environmental Impact Report Vol. 1.*

LADWP and Ecosystem Sciences. 2010. "Owens Valley Land Management Plan." https://www.inyowater.org/wp-content/uploads/2013/11/Owens-Valley-Land-Management-Plan-Final.pdf.

Lancaster, Nicholas, and Steven N. Bacon. 2012. *Late Holocene Stratigraphy and Chronology of Keeler Dunes Area.* Report prepared by Desert Research Institute for Great Basin Unified Air Pollution Control District.

Lancaster, Nicholas, and Grace McCarley-Holder. 2013. "Decadal-Scale Evolution of a Small Dune Field: Keeler Dunes, California 1944–2010." *Geomorphology* 180–181:281–291. https://doi.org/10.1016/j.geomorph.2012.10.017.

Lancaster, Nicholas, Sophie Baker, Steven N. Bacon, and Grace McCarley-Holder. 2015. "Owens Lake Dune Fields: Composition, Sources of Sand, and Transport Pathways." *CATENA* 134:41–49. https://doi.org/10.1016/j.catena.2015.01.003.

Lawton, Harry W., Philip J. Wilke, Mary DeDecker, and William M. Mason. 1976. "Agriculture Among the Paiute of Owens Valley." *The Journal of California Anthropology* 3 (1):13–50.

Lee, Charles Hamilton. 1915. *Report on Hydrology of Owens Lake Basin and the Natural Soda Industry as Affected by the Los Angeles Aqueduct Diversion.* Los Angeles Bureau of Water Works & Supply.

Li, Hong-Chun, James L. Bischoff, Teh-Lung Ku, Steven P. Lund, and Lowell D. Stott. 2000. "Climate Variability in East-Central California During the Past 1000 Years Reflected by High-Resolution Geochemical and Isotopic Records from Owens Lake Sediments." *Quaternary Research* 54 (2):189–197.

Li, Junran, Gregory S. Okin, Jeffrey E. Herrick, Jayne Belnap, Mark E. Miller, Kimberly Vest, and Amy E. Draut. 2013. "Evaluation of a New Model of Aeolian Transport in the Presence of Vegetation." *Journal of Geophysical Research F: Earth Surface* 118 (1):288–306. https://doi.org/10.1002/jgrf.20040.

Li, Junran, Sujith Ravi, Guan Wang, R. Scott Van Pelt, Thomas E. Gill, and Joel B. Sankey. 2022. "Woody Plant Encroachment of Grassland and the Reversibility of Shrub Dominance: Erosion, Fire, and Feedback Processes." *Ecosphere* 13 (3):e3949.

Li, Shuanhu, Chi Li, De Yao, and Shuo Wang. 2020. "Feasibility of Microbially Induced Carbonate Precipitation and Straw Checkerboard Barriers on Desertification Control and Ecological Restoration." *Ecological Engineering* 152:105883.

Li, X. R., H. L. Xiao, M. Z. He, and J. G. Zhang. 2006. "Sand Barriers of Straw Checkerboards for Habitat Restoration in Extremely Arid Desert Regions." *Ecological Engineering* 28 (2):149–157.

Liu, Yang, and David J. Diner. 2017. "Multi-Angle Imager for Aerosols." *Public Health Reports* 132 (1):14–17. https://doi.org/10.1177/0033354916679983.

Lihui, Tian, Wu Wangyang, Zhang Dengshan, Lu Ruijie, and Wang Xuequan. 2015. "Characteristics of Erosion and Deposition of Straw Checkerboard Barriers in Alpine Sandy Land." *Environmental Earth Sciences* 74:573–584.

Lu, Hua, and Yaping Shao. 1999. "A New Model for Dust Emission by Saltation Bombardment." *Journal of Geophysical Research: Atmospheres* 104 (D14):16827–16842.

Lu, Ning, and William J. Likos. 2004. *Unsaturated Soil Mechanics.* Hoboken, NJ: J. Wiley.

Lyu, Yanli, Peijun Shi, Guoyi Han, Lianyou Liu, Lanlan Guo, Xia Hu, and Guoming Zhang. 2020. "Desertification Control Practices in China." *Sustainability* 12 (8):3258.

Madley, Benjamin. 2016. *An American Genocide: The United States and the California Indian Catastrophe, 1846–1873.* Yale University Press.

Manning, Sara J. 1997. "Plant Communities of LADWP Land in the Owens Valley: An Explanatory Analysis of Baseline Conditions." *Inyo County, California: Inyo County Water Department*:161.

Manning, Sara J. 1999. "The Effects of Water Table Decline on Groundwater-Dependent Great Basin Plant Communities in the Owens Valley, California." Shrubland Ecotones, Ephraim, Utah.

Mayaud, Jerome R., Richard M. Bailey, and Giles F. S. Wiggs. 2017. "A Coupled Vegetation/Sediment Transport Model for Dryland Environments." *Journal of Geophysical Research: Earth Surface* 122 (4):875–900.

McLendon, Terry., Elke Naumburg, and David W. Martin. 2012. "Secondary Succession Following Cultivation in an Arid Ecosystem: The Owens Valley, California." *Journal of Arid Environments* 82:136–146.

Mejia, John F., John A. Gillies, Vicken Etyemezian, and R. Glick. 2019. "A Very-High Resolution (20m) Measurement-Based Dust Emissions and Dispersion Modeling Approach for the Oceano Dunes, California." *Atmospheric Environment* 218.

Mekdaschi, Rima, and Hans Peter Liniger. 2013. *Water Harvesting: Guidelines to Good Practice.* Centre for Development and Environment.

Meyers, Zachary P., Marty D. Frisbee, Laura K. Rademacher, and Noah S. Stewart-Maddox. 2021. "Old Groundwater Buffers the Effects of a Major Drought in Groundwater-Dependent Ecosystems of the Eastern Sierra Nevada (CA)." *Environmental Research Letters* 16 (4):044044.

Mihevc, Todd M., Gilbert F. Cochran, and Mary Feeney Hall. 1997. *Simulation of Owens Lake Water Levels*: Water Resources Center, Desert Research Institute, University and Community College System of Nevada.

Milankovitch, Milutin K. 1941. *Kanon der Erdbestrahlung und seine Anwendung auf das Eiszeitenproblem.* Royal Serbian Academy Special Publication 133.

Molina Rueda, Emilio, Ellison Carter, Christian L'Orange, Casey Quinn, and John Volckens. 2023. "Size-Resolved Field Performance of Low-Cost Sensors for Particulate Matter Air Pollution." *Environmental Science & Technology Letters* 10 (3):247–253. https://doi.org/10.1021/acs.estlett.3c00030.

Muhs, Daniel R. 2004. "Mineralogical Maturity in Dunefields of North America, Africa and Australia." *Geomorphology* 59 (1):247–269. https://doi.org/10.1016/j.geomorph.2003.07.020.

Napelenok, S. L., Robert W. Pinder, Alice B. Gilliland, and Randall V. Martin. 2008. "A Method for Evaluating Spatially-Resolved NO_x Emissions Using Kalman Filter Inversion, Direct Sensitivities, and Space-Based NO_2 Observations." *Atmospheric Chemistry and Physics* 8 (18):5603–5614.

NASEM (National Academies of Sciences, Engineering, and Medicine). 2020. *Effectiveness and Impacts of Dust Control Measures for Owens Lake.* Washington, DC: The National Academies Press.

NASA (National Aeronautics and Space Administration). 2024. "The MAIA Mission." https://maia.jpl.nasa.gov/mission.

NASA. n.d. "EMIT Earth Surface Mineral Dust Source Investigation." https://earth.jpl.nasa.gov/emit/mission/about.

National Centers for Environmental Information. 2025. "U.S. Monthly Climate Normals (1981–2010)." National Oceanic and Atmospheric Administration. https://www.ncei.noaa.gov/access/metadata/landing-page/bin/iso?id=gov.noaa.ncdc:C00822.

Nichols, William D. 1994. "Groundwater Discharge by Phreatophyte Shrubs in the Great Basin as Related to Depth to Groundwater." *Water Resources Research* 30 (12):3265–3274.

NOAA (National Oceanic and Atmospheric Administration). 2024a. "VIIRS Imagery." https://vlab.noaa.gov/web/towr-s/viirs-img.

NOAA. 2024b. "Visible Infrared Imaging Radiometer Suite (VIIRS)." https://www.nesdis.noaa.gov/our-satellites/currently-flying/joint-polar-satellite-system/visible-infrared-imaging-radiometer-suite-viirs.

O'Brian, Jon. 2021. *Dust Emissions and OHV Activity at ODSVRA - Update on the Oceano Dust Program and Recent Research.* Sacramento, CA: California Natural Resources Agency.

Okin, Gregory S. 2008. "A New Model of Wind Erosion in The Presence of Vegetation." *Journal of Geophysical Research: Earth Surface* 113 (F2). https://doi.org/10.1029/2007JF000758.

Okin, Gregory S., Dale A. Gillette, and Jeffrey E. Herrick. 2006. "Multi-Scale Controls on and Consequences of Aeolian Processes in Landscape Change in Arid and Semi-Arid Environments." *Journal of Arid Environments* 65 (2):253–275.

Okin, Gregory S., Anthony J. Parsons, John Wainwright, Jeffrey E. Herrick, Brandon T. Bestelmeyer, Debra C. Peters, and Ed L. Fredrickson. 2009. "Do Changes in Connectivity Explain Desertification?" *BioScience* 59 (3):237–244.

Okin, Gregory S., Mariano Moreno-de las Heras, Patricia M. Saco, Heather L. Throop, Enrique R. Vivoni, Anthony J. Parsons, John Wainwright, and Debra P.C. Peters. 2015. "Connectivity in Dryland Landscapes: Shifting Concepts of Spatial Interactions." *Frontiers in Ecology and the Environment* 13 (1):20–27.

Ouimette, James R., William C. Malm, Bret A. Schichtel, Patrick J. Sheridan, Elisabeth Andrews, John A. Ogren, and W. Patrick Arnott. 2022. "Evaluating the PurpleAir Monitor as an Aerosol Light Scattering Instrument." *Atmospheric Measurement Techniques* 15 (3):655–676. https://doi.org/10.5194/amt-15-655-2022.

Ono, Duane. 2006. "Application of the Gillette Model for Windblown Dust at Owens Lake, CA." *Atmospheric Environment* 40 (17):3011–3021. https://doi.org/10.1016/j.atmosenv.2005.08.048.

Ono, Duane, and Chris Howard. 2016. *Appendix G, GBUAPCD Off-lake PM_{10} Reductions Memorandum, Bishop, CA.* Bishop, California: Great Basin Unified Air Pollution Control District.

Ono, Duane, Phill Kiddoo, Christopher Howard, Guy Davis, and Kenneth Richmond. 2011. "Application of a Combined Measurement and Modeling Method to Quantify Windblown Dust Emissions from the Exposed Playa at Mono Lake, California." *Journal of the Air & Waste Management Association* 61 (10): 1036–1045. https://doi.org/10.1080/10473289.2011.596760.

Owens Valley Groundwater Authority. 2021. "Owens Valley Basin Final Groundwater Sustainability Plan." https://ovga.us/wp-content/uploads/2021/12/OVGA_groundwater_sustainability_plan_Final-120921.pdf.

Owens Valley Groundwater Authority. 2024. "Owens Valley and Fish Slough Subbasins GSP Annual Report Water Year 2022." https://ovga.us/wp-content/uploads/2024/07/OVGA_Annual_Report_WY2022_FINAL_rev_signed_R.pdf.

Owens Valley Indian Water Commission. 2024. "Owens Valley." https://www.oviwc.org/owens-valley.

Oxford Reference. 2015. "Histic Epipedon" *A Dictionary of Ecology*, 5th edition.

Özkaynak, Halûk, Lisa K. Baxter, Kathie L. Dionisio, and Janet Burke. 2013. "Air Pollution Exposure Prediction Approaches Used in Air Pollution Epidemiology Studies." *Journal of Exposure Science & Environmental Epidemiology* 23 (6):566–572.

Paatero, Pentti, and Unto Tapper. 1994. "Positive Matrix Factorization: A Non-Negative Factor Model with Optimal Utilization of Error Estimates of Data Values." *Environmetrics* 5 (2):111–126.

REFERENCES

Parajuli, Sagar P., and Charles S. Zender. 2018. "Projected Changes in Dust Emissions and Regional Air Quality Due to the Shrinking Salton Sea." *Aeolian Research* 33:82–92.

Pennington, Elyse A., Yuan Wang, Benjamin C. Schulze, Karl M. Seltzer, Jiani Yang, Bin Zhao, Zhe Jiang, Hongru Shi, Melissa Venecek, and Daniel Chau. 2024. "An Updated Modeling Framework to Simulate Los Angeles Air Quality–Part 1: Model Development, Evaluation, and Source Apportionment." *Atmospheric Chemistry and Physics* 24 (4):2345–2363.

Perkins, Kim S., David M. Miller, Darren R. Sandquist, Miguel Macias, and Aimee Roach. 2018. "Ecohydrologic Changes Caused by Hydrologic Disconnection of Ephemeral Stream Channels in Mojave National Preserve, California." *Vadose Zone Journal* 17 (1):1–8.

Perry, Steven G., Alan J. Cimorelli, Robert J. Paine, Roger W. Brode, Jeffrey C. Weil, Akula Venkatram, Robert B. Wilson, Russell F. Lee, and Warren D. Peters. 2005. "AERMOD: A Dispersion Model for Industrial Source Applications. Part II: Model Performance Against 17 Field Study Databases." *Journal of Applied Meteorology* 44 (5):694–708.

Peters, Debra P. C., Gregory S. Okin, Jeffrey E. Herrick, Heather M. Savoy, John P. Anderson, Stacey L. P. Scroggs, and Junzhe Zhang. 2020. "Modifying Connectivity to Promote State Change Reversal: The Importance of Geomorphic Context and Plant–Soil Feedbacks." *Ecology* 101 (9):e03069.

Pigati, Jeffrey S., Kathleen B. Springer, Harrison J. Gray, Matthew R. Bennett, and David Bustos. 2024. "The Geochronology of White Sands Locality 2 is Resolved." *PaleoAmerica* 10 (1):28–44.

Polade, Suraj. D., Alexander Gershunov, Daniel R. Cayan, Michael D. Dettinger, and David W. Pierce. 2017. "Precipitation in a Warming World: Assessing Projected Hydro-Climate Changes in California and Other Mediterranean Climate Regions." *Scientific Reports* 7. https://doi.org/10.1038/s41598-017-11285-y.

Potter, K. N., and T. M. Zobeck. 1990. "Estimation of Soil Microrelief." *Transactions of the ASAE* 33 (1):156–161. https:/doi.org/10.13031/2013.31310.

Presley, DeAnn, and John Tatarko. 2009. *Principles of Wind Erosion and its Control*. Kansas State University.

Pritchett, Daniel W., and Sara J. Manning. 2009. "Effects of Fire and Groundwater Extraction on Alkali Meadow Habitat in Owens Valley, California." *Madroño* 56 (2):89–98.

Qiu, Guo Yu, In-Bok Lee, Hideyuki Shimizu, Yong Gao, and Guodong Ding. 2004. "Principles of Sand Dune Fixation with Straw Checkerboard Technology and its Effects on the Environment." *Journal of Arid Environments* 56 (3):449–464.

Qu, Zhen, Daven K. Henze, Helen M. Worden, Zhe Jiang, Benjamin Gaubert, Nicolas Theys, and Wei Wang. 2022. "Sector-Based Top-Down Estimates of NO, SO2, and CO Emissions in East Asia." *Geophysical Research Letters* 49 (2):e2021GL096009. https://doi.org/10.1029/2021GL096009.

Quick, Dayna J., and Oliver A. Chadwick. 2011. "Accumulation of Salt-Rich Dust from Owens Lake Playa in Nearby Alluvial Soils." *Aeolian Research* 3 (1):23–29.

Rachal, David M., Gregory S. Okin, C. Alexander, Jeffrey E. Herrick, and Debra P. C. Peters. 2015. "Modifying Landscape Connectivity by Reducing Wind Driven Sediment Redistribution, Northern Chihuahuan Desert, USA." *Aeolian Research* 17:129–137.

Ramboll Americas Engineering Solutions, Inc. 2020. *User's Guide Comprehensive Air Quality Model with Extensions*. https://www.camx.com/Files/CAMxUsersGuide_v7.30.pdf

Raupach, Michael R. 1992. "Drag and Drag Partition on Rough Surfaces." *Boundary-Layer Meteorology* 60 (4):375–395.

Raupach, Michael R., Ngaire Woods, Gary Dorr, John F. Leys, and Helen A. Cleugh. 2001. "The Entrapment of Particles by Windbreaks." *Atmospheric Environment* 35 (20):3373–3383. https://doi.org/10.1016/S1352-2310(01)00139-X.

Reheis, Marith C. 1997. "Dust Deposition Downwind of Owens (Dry) Lake, 1991–1994: Preliminary Findings." *Journal of Geophysical Research: Atmospheres* 102 (D22):25999–26008. https://doi.org/10.1029/97JD01967.

Reheis, Marith C., James R. Budahn, and Paul J. Lamothe. 2002. "Geochemical Evidence for Diversity of Dust Sources in the Southwestern United States." *Geochimica et Cosmochimica Acta* 66 (9):1569–1587.

Richards, James H., Jason K. Smesrud, Dane L. Williams, Brian M. Schmid, John B. Dickey, and Maarten D. Schreuder. 2022. "Vegetation, Hydrology, and Sand Movement Interactions on the Slate Canyon Alluvial Fan-Keeler Dunes Complex, Owens Valley, California." *Aeolian Research* 54:100773.

Richmond, Ben. 2019. "Beyond the Exceptional Events Rule." *Ecology Law Quarterly* 46 (2):343–372.

Riter, Karmann, Prakash Doraiswamy, Anthony Clint Clayton, and Kelley Rountree. 2023. *Solar Station for an Off-The-Grid Air Quality Sensor System*. RTI Press Publication No. MR-0051-2306. Research Triangle Park, NC: RTI Press. https://doi.org/10.3768/rtipress.2023.mr.0051.2306

Rodriguez-Caballero, Emilio, Tanja Stanelle, Sabine Egerer, Yafang Cheng, Hang Su, Y. Canton, Jayne Belnap, Meinrat O. Andreae, Ina Tegen, and Christian H. Reick. 2022. "Global Cycling and Climate Effects of Aeolian Dust Controlled by Biological Soil Crusts." *Nature Geoscience* 15 (6):458–463.

Rosenthal, Jeffrey S., Jack Meyer, Manuel R. Palacios-Fest, D. Craig Young, Andrew Ugan, Brian F. Byrd, Ken Gobalet, and Jason Giacomo. 2017. "Paleohydrology of China Lake Basin and the Context of Early Human Occupation in the Northwestern Mojave Desert, USA." *Quaternary Science Reviews* 167:112–139.

Sahagun, Louis. 2024. "Cost of Owens Valley storm damage continues to mount for Los Angeles Department of Water and Power." *LA Times* February 18.

Saint-Armand, Pierre, Larry A. Mathews, Camille Gaines, and Roger Reinking. 1986. *Dust Storms from Owens and Mono Valleys, California*. NWC TP 6731. Naval Weapons Center, China Lake, CA.

Salton Sea Management Program. 2024. *Salton Sea MIP Annual Work Plan*. https://saltonsea.ca.gov/wp-content/uploads/2024/04/2024-Salton-MIP-Work-Plan_ADA-Final.pdf.

Schlesinger, William H., and Cynthia S. Jones. 1984. "The Comparative Importance of Overland Runoff and Mean Annual Rainfall to Shrub Communities of the Mojave Desert." *Botanical Gazette* 145(1):116–124.

Schmid, Brian M., Dane L. Williams, Chuan-Shin Chong, Miles D. Kenney, John B. Dickey, and Peter Ashley. 2022. "Use of Digital Photogrammetry and LiDAR Techniques to Quantify Time-Series Dune Volume Estimates of the Keeler Dunes Complex, Owens Valley, California." *Aeolian Research* 54:100764.

Schweizer, Donald, Ricardo Cisneros, and Monica Buhler. 2019. "Coarse and fine particulate matter components of wildland fire smoke at Devils Postpile National Monument, California, USA." *Aerosol and Air Quality Research* 19(7): 1463–1470.

Scire, Joseph S., Francoise R. Robe, Mark E. Fernau, and Robert J. Yamartino. 2000. "A User's Guide for the CALMET Meteorological Model." *Earth Tech, USA* 37.

Scire, Joseph S., David G. Strimaitis, and Robert J.Yamartino. 2000. "A User's Guide for the CALPUFF Dispersion Model." *Earth Tech, Inc* 521.

Sehmel, George A. 1980. "Particle and Gas Dry Deposition: A Review." *Atmospheric Environment* 14 (9):983–1011.

Serafin, Stefano, Lukas Strauss, and Vanda Grubišić. 2017. "Climatology of Westerly Wind Events in the Lee of the Sierra Nevada." *Journal of Applied Meteorology and Climatology* 56 (4):1003–1023.

Shao, Yaping, M. R. Raupach, and P. A. Findlater. 1993. "Effect of Saltation Bombardment on the Entrainment of Dust by Wind." *Journal of Geophysical Research: Atmospheres* 98 (D7):12719–12726.

Sherman, Douglas J., and J. T. Ellis. 2022. "Sand Transport Processes." In *Treatise on Geomorphology*, edited by Nicholas Lancaster, 385–414. Amsterdam: Elsevier Inc.

Skamarock, William C., Joseph B. Klemp, Jimy Dudhia, David O. Gill, Dale M. Barker, Wei Wang, and Jordan G. Powers. 2007. *A Description of the Advanced Research WRF Version 2*. National Center for Atmospheric Research.

Smith, George I., James L. Bischoff, and J. Platt Bradbury. 1997. "Synthesis of the Paleoclimatic Record from Owens Lake Core OL-92." https://doi.org/10.1130/0-8137-2317-5.143.

Smoot, J. P., R. J. Litwin, J. L. Bischoff, and S. J. Lund. 2000. "Sedimentary Record of the 1872 Earthquake and 'Tsunami' at Owens Lake, Southeast California." *Sedimentary Geology* 135 (1-4):241–254.

South Coast Air Quality Management District. 2024. "Summary Table and Reports for PM Sensors." http://www.aqmd.gov/aq-spec/evaluations/criteria-pollutants/summary-pm.

State of California Department of Parks and Recreation. 2024. "Oceano Dunes State Vehicular Recreation Area Dust Control Program Provisional Final 2024 Annual Report and Work Plan." https://mesaairfacts.net/index_htm_files/2024ARWP_ProvFinal_20240911_reduced.pdf.

Stavrakou, T., and J.-F. Müller. 2006. "Grid-Based Versus Big Region Approach for Inverting CO Emissions Using Measurement of Pollution in the Troposphere (MOPITT) Data." *Journal of Geophysical Research: Atmospheres* 111 (D15304).

Steward, Julian Haynes. 1933. *Ethnography of the Owens Valley Paiute*. University of California Publications in American Archaeology and Ethnology 33. Berkeley, CA: University of California Press.

Stillwater Sciences. 2021. *Assessment of Groundwater Dependent Ecosystems for the Owens Valley Basin Groundwater Stability Plan Technical Appendix*. https://ovga.us/wp-content/uploads/2021/12/Appendix-9-GDE-OVGA-Final-GSP-.pdf

Stone, Paul, George C. Dunne, James G. Moore, and George I. Smith. 2000. *Geologic Map of the Lone Pine 15'Quadrangle, Inyo County, California*. Reston, VA: U. S. Geological Survey.

Swain, Daniel L., Daniel E. Horton, Deepti Singh, and Noah S. Diffenbaugh. 2016. "Trends in Atmospheric Patterns Conducive to Seasonal Precipitation and Temperature Extremes in California." *Science Advances* 2 (4):e1501344. https://doi.org/10.1126/sciadv.1501344.

Swain, Daniel L., Baird Langenbrunner, J. David Neelin, and Alex Hall. 2018. "Increasing Precipitation Volatility in Twenty-First-Century California." *Nature Climate Change* 8 (5):427–433. https://doi.org/10.1038/s41558-018-0140-y.

Sweeney, Mark R. 2022. "Dust Emission Processes." In *Treatise on Geomorphology*, edited by J. J. F. Shroder, 235–258. Amsterdam: Elsevier.

REFERENCES

Sweeney, Mark R., Tad Lacey, and Steven L. Forman. 2023. "The Role of Abrasion and Resident Fines in Dust Production from Aeolian Sands as Measured by the Portable In-Situ Wind Erosion Laboratory (PI-SWERL)." *Aeolian Research* 63:100889.

Tatarko, John, Wilma Trujillo, and Meagan Schipansk. 2019. *Wind Erosion Processes and Control*. Colorado State University Extention.

Tsoar, Haim, Noam Levin, Naomi Porat, Luis P. Maia, Hans J. Herrmann, Sonia H. Tatumi, and Vanda Claudino-Sales. 2009. "The Effect of Climate Change on the Mobility and Stability of Coastal Sand Dunes in Ceará State (NE Brazil)." *Quaternary Research* 71(2):217–226.

Tyler, S. W., S. Kranz, M. B. Parlange, J. Albertson, G. Cochran, B. Lyles, and G. Holder. 1997. "Estimation of Groundwater Evaporation and Salt Flux from Owens Lake, California, USA." *Journal of Hydrology* 200:110–135.

USDA (United States Department of Agriculture). 2016. "Official Soil Series Descriptions." https://www.nrcs.usda.gov/resources/data-and-reports/official-soil-series-descriptions-osd.

USGS (United States Geological Survey). 2024. "Sand Sheet." https://apps.usgs.gov/thesaurus/term-simple.php?thcode=4&code=1.3.2.

Van Pelt, R. Scott, John Tatarko, Thomas E. Gill, Chunping Chang, Junran Li, Iyasu G. Eibedingil, and Marcos Mendez. 2020. "Dust Emission Source Characterization for Visibility Hazard Assessment on Lordsburg Playa in Southwestern New Mexico, USA." *Geoenvironmental Disasters* 7(1):34. https://doi.org/10.1186/s40677-020-00171-x.

Varner, Gary R. 2009. *The Owens Valley Paiute: A Cultural History*. Morrisville, NC: Lulu Press Inc.

Vautard, Robert, Pascal Yiou, and Michael Ghil. 1992. "Singular-Spectrum Analysis: A Toolkit for Short, Noisy Chaotic Signals." *Physica D: Nonlinear Phenomena* 58 (1–4):95–126.

Venkatram, Akula, and Alan J. Cimorelli. 2007. "On the Role of Nighttime Meteorology in Modeling Dispersion of Near Surface Emissions in Urban Areas." *Atmospheric Environment* 41(4):692–704.

Vest, Kimberly R., Andrew J. Elmore, James M. Kaste, Gregory S. Okin, and Junran Li. 2013. "Estimating Total Horizontal Aeolian Flux Within Shrub-Invaded Groundwater-Dependent Meadows Using Empirical And Mechanistic Models." *Journal of Geophysical Research: Earth Surface* 118 (2):1132–1146. https://doi.org/10.1002/jgrf.20048.

Vohra, Karn, Alina Vodonos, Joel Schwartz, Eloise A. Marais, Melissa P. Sulprizio, and Loretta J. Mickley. 2021. "Global Mortality from Outdoor Fine Particle Pollution Generated by Fossil Fuel Combustion: Results from GEOS-Chem." *Environmental Research* 195.

Wacaser, R., D. James, H. Jeong, and S. Pulurgurtha. 2006. "Refined PM_{10} Aeolian Emission Factors for Native Desert and Disturbed Vacant Areas." Appendix E of *PM_{10} State Implementation Milestone Achievement Report*. Los Vegas: Nevada Department of Air Quality and Environment.

Walker, Ian J., Zach Hilgendorf, John A. Gillies, Craig M. Turner, Eden Furtak-Cole, and George Nikolich. 2023. "Assessing Performance of a "Nature-Based" Foredune Restoration Project, Oceano Dunes, California, USA." *Earth Surface Processes and Landforms* 48(1):143–162. https://doi.org/10.1002/esp.5478.

Wang, Juan, and Rui Wang. 2019. "The Physical and Chemical Properties of Soil Crust in Straw Checkerboards with Different Ages in the Mu Us sandland, Northern China." *Sustainability* 11(17):4755.

Wang, Jun, Xiaoguang Xu, Daven K. Henze, Jing Zeng, Qiang Ji, Si-Chee Tsay, and Jianping Huang. 2012. "Top-down Estimate of Dust Emissions Through Integration of MODIS and MISR Aerosol Retrievals with the GEOS-Chem Adjoint Model." *Geophysical Research Letters* 39(8). https://doi.org/10.1029/2012GL051136.

Wang, Tao, Jianjun Qu, and Qinghe Niu. 2020. "Comparative Study of the Shelter Efficacy of Straw Checkerboard Barriers and Rocky Checkerboard Barriers in a Wind Tunnel." *Aeolian Research* 43:100575. 10.1016/j.aeolia.2020.100575.

Wang, Xiaoliang, John A. Gillies, Steven Kohl, Eden Furtak-Cole, Karl A. Tupper, and David A. Cardiel. 2023. "Quantifying the Source Attribution of PM_{10} Measured Downwind of the Oceano Dunes State Vehicular Recreation Area." *Atmosphere* 14(4):718.

Watson, John G. 1984. "Overview of Receptor Model Principles." *Journal of the Air Pollution Control Association* 34(6):619–623.

Watson, John G., John A. Cooper, and James J. Huntzicker. 1984. "The Effective Variance Weighting for Least Squares Calculations Applied to the Mass Balance Receptor Model." *Atmospheric Environment* 18(7):1347–1355.

Webb, Nicholas P., and Caroline Pierre. 2018. "Quantifying Anthropogenic Dust Emissions." *Earth's Future* 6(2):286–295.

Wheaton, Joseph M., James Brasington, Stephen E. Darby, and David A. Sear. 2010. "Accounting for Uncertainty in DEMs from Repeat Topographic Surveys: Improved Sediment Budgets." *Earth Surface Processes and Landforms: The Journal of the British Geomorphological Research Group* 35 (2):136–156.

Williams, A. Park, John T. Abatzoglou, Alexander Gershunov, Janin Guzman-Morales, Daniel A. Bishop, Jennifer K. Balch, and Dennis P. Lettenmaier. 2019. "Observed Impacts of Anthropogenic Climate Change on Wildfire in California." *Earth's Future* 7 (8):892–910.

Xin, Guowei, Ning Huang, Jie Zhang, and Hongchao Dun. 2021. "Investigations into the Design of Sand Control Fence for Gobi Buildings." *Aeolian Research* 49:100662.

Xu, Bin, Jie Zhang, Ning Huang, Kang Gong, and Yusheng Liu. 2018. "Characteristics of Turbulent Aeolian Sand Movement Over Straw Checkerboard Barriers and Formation Mechanisms of Their Internal Erosion Form." *Journal of Geophysical Research: Atmospheres* 123 (13):6907–6919. https://doi.org/10.1029/2017JD027786.

Yizhaq, Hezi, Y. Ashkenazy, and H. Tsoar. 2009. "Sand dune dynamics and climate change: A Modeling Approach." *Journal of Geophysical Research: Earth Surface* 114 (F1). https://doi.org/10.1029/2008JF001138.

Yizhaq, Hezi, Yosef Ashkenazy, and Haim Tsoar. 2007. "Why Do Active and Stabilized Dunes Coexist under the Same Climatic Conditions?" *Physical Review Letters* 98 (18):188001. https://doi.org/10.1103/PhysRevLett.98.188001.

Zhang, Chun-Lai, Xue-Yong Zou, Hong Cheng, Shuo Yang, Xing-Hui Pan, Yu-Zhang Liu, and Guang-Rong Dong. 2007. "Engineering Measures to Control Windblown Sand in Shiquanhe Town, Tibet." *Journal of Wind Engineering and Industrial Aerodynamics* 95:53-70. https://doi.org/10.1016/j.jweia.2006.05.006.

Zhang, Lin, Daniel J. Jacob, Monika Kopacz, Daven K. Henze, Kumaresh Singh, and Daniel A. Jaffe. 2009. "Intercontinental Source Attribution of Ozone Pollution at Western US Sites Using an Adjoint Method." *Geophysical Research Letters* 36 (L11810).

Zhang, Shuai, Guo-dong Ding, Ming-han Yu, Guang-lei Gao, Yuan-yuan Zhao, Guo-hong Wu, and Long Wang. 2018. "Effect of Straw Checkerboards on Wind Proofing, Sand Fixation, and Ecological Restoration in Shifting Sandy Land." *International Journal of Environmental Research and Public Health* 15 (10):2184.

Zhang, Weimin, Tao Wang, Wangfu Wang, Jianjun Qu, Xian Xue, and Zhengyi Yao. 2004. "The Gobi Sand Stream and its Control over the Top Surface of the Mogao Grottoes, China." *Bulletin of Engineering Geology and the Environment* 63 (3):261–269. https://doi.org/10.1007/s10064-004-0239-4.

Zhang, Weixin, Da Pan, I-Ting Ku, and Jeffrey L. Collett Jr. 2024. "Quantification of Volatile Organic Compound Emissions from Unconventional Oil and Gas Development [Dataset]." *Dryad.* https://doi.org/10.5061/dryad.g1jwstqzs.

Zhao, Wenzhi, Guanglu Hu, Zhihui Zhang, and Zhibin He. 2008. "Shielding Effect of Oasis-Protection Systems Composed of Various Forms of Wind Break on Sand Fixation in an Arid Region: A Case Study in the Hexi Corridor, Northwest China." *Ecological Engineering* 33 (2):119–125. https://doi.org/10.1016/j.ecoleng.2008.02.010.

Zheng, Tongshu, Michael H. Bergin, Shijia Hu, Joshua Miller, and David E. Carlson. 2020. "Estimating Ground-level PM2.5 Using Micro-Satellite Images by a Convolutional Neural Network and Random Forest Approach." *Atmospheric Environment* 230:117451.

Zhong, Shiyuan, Ju Li, C. David Whiteman, Xindi Bian, and Wenqing Yao. 2008. "Climatology of High Wind Events in the Owens Valley, California." *Monthly Weather Review* 136 (9):3536–3552.

Zhou, Xiaobing, Yunge Zhao, Jayne Belnap, Bingchang Zhang, Chongfeng Bu, and Yuanming Zhang. 2020. "Practices of Biological Soil Crust Rehabilitation in China: Experiences and Challenges." *Restoration Ecology* 28 (S2):S45–S55. https://doi.org/10.1111/rec.13148.

Zobeck, Ted M., and R. Scott Van Pelt. 2011. "Wind Erosion." *Soil Management: Building a Stable Base for Agriculture*: 209–227.

Zucca, Claudio, Renate Fleiner, Enrico Bonaiuti, and Utchang Kang. 2022. "Land Degradation Drivers of Anthropogenic Sand and Dust Storms." *Catena* 219:106575.

Appendix A

Emission Fluxes of On- and Off-Lake Sources

This appendix explains the approach used to compare emission fluxes between on- and off-lake sources impacting monitors around Owens Lake adjusted for meteorology. This analysis is used to assess the relative level of emissions leading to exceedances, potential for future exceedances related to historical emissions, and the relative emission impacts from on- versus off-lake sources. More detailed modeling (e.g., using dispersion modeling) could be used to better quantify emissions from specific source areas, more fully accounting for source location and system geometry, as recommended in the report.

EMISSION FLUXES OF ON- AND OFF-LAKE SOURCES

The figures in Chapter 3 (Figures 3-5, 3-8, 3-11, 3-14, 3-17, and 3-21) show that most of the PM_{10} monitoring stations are surrounded by on- and off-lake dust sources. This suggests the application of an idea that underlies a dispersion model proposed by Gifford and Hanna (1970) and Hanna (1971) to compute pollutant concentrations in an urban area. The model assumes that a monitor measuring concentrations is surrounded by areas with pollutant emissions that are relatively uniform across an urban area. Gifford and Hanna (1970) then show that the seasonally averaged concentration at the monitor is governed primarily by the emission flux (emission rate per unit area of the source) of the upwind area in the immediate vicinity of the monitor and the wind speed at the monitor. The concentration, at the monitor is given by the simple formula

$$C = A\frac{q}{U} \qquad (1)$$

where q is the emission flux of the upwind area contributing to the concentration, and U is the wind speed at the monitor. The variable $A = CU/q$, derived from annual averages of particulate concentrations (total suspended particles) and $1/U$, is relatively insensitive to the size of the city (or contributing area) as seen in Figure A-1. A varies by a factor of two around the value of 225 when the inferred size of the city (or contributing area) varies by a factor of 7. This indicates that the product of the PM_{10} concentration, C, and the wind speed at a monitor, U, can be used to compare the relative magnitudes of upwind emission fluxes at different monitors as long as the extents of upwind areas affecting the monitors do not differ significantly.

The locations of the monitors around Owens Lake relative to the sources impacting them are similar to those in the urban areas studied by Gifford and Hanna (1973) in that they are surrounded by sources of dust. This sug-

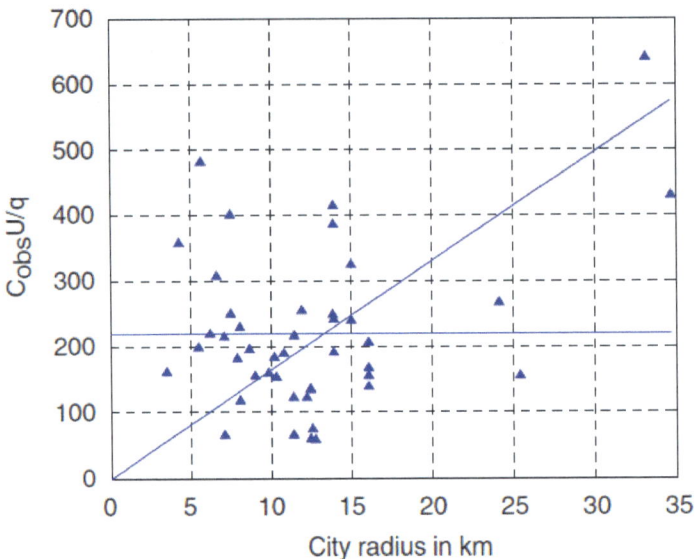

FIGURE A-1 CU/q is relatively insensitive compared to the size of the city.
NOTES: The sloped blue line is a trend line indicating that the ratio CU/q is insensitive to city size. The horizontal blue line is the mean value suggested by Gifford and Hanna (1973).
SOURCE: Based on data from Venkatram and Cimorelli (2007).

gests a method to estimate fluxes from the upwind areas of the monitors using the relationship $q = CU/A$. The relative insensitivity of A to the extent of the upwind area that impacts a monitor can be readily confirmed with a calculation that accounts for the heights of the PM_{10} monitors and wind measurements and the neutral stability associated with the wind speeds that gives rise to exceedances. Furthermore, the magnitude of A does not play a role in assessing emissions trends at a site if the source-receptor geometry does not change over time (i.e., the source area and monitor location are consistent).

Using hourly wind speed, wind direction, and PM_{10} concentration data from all monitors from 2000 to 2023 (Chris Howard, personal communication, November 2024; Ann Logan, personal communication, July 2024), the panel calculated normalized emission fluxes for monitors around the lake. The wind directions used in computing UC correspond to the wind sectors that the Greater Basin Unified Air Pollution Control District (the District) associates with on- and off-lake dust sources. The emission flux averaged over a year is $q(average) = average(U_i C_i)$, where the subscript '$i$' corresponds to hourly concentrations and wind speeds measured from the on- and off-lake sectors over 1 year, and the factor A is absorbed in the normalization, described below. Given the sensitivity of PM_{10} emissions to wind speed (Gillette, Ono, and Richmond 2004; Jiang, Liu, and Doyle 2011), effective emission fluxes are taken to be averages over values corresponding to wind speeds between 10 and 20 m/s, a range likely to yield peak monitor-recorded concentrations. This range is similar to the maximum average hourly wind speeds observed at Owens Lake during exceedance events (see Figure 5-2).

The annual computed values of q at the monitors were normalized by the value of q at Keeler averaged over the entire period (thus, on the graph for Keeler, the normalized emissions cross one during the period); this provides a comparison of emission fluxes relative to the value at Keeler. The choice of Keeler is made because of the relatively large number of exceedances in Keeler that have been attributed to off-lake sources. Normalizing by results from another location would only shift the values found and not change the findings.

The magnitudes of the normalized emission fluxes can be interpreted by plotting them against the associated annual exceedances at the monitors (Figure A-2); these exceedances are related to on- and off-lake dust sources

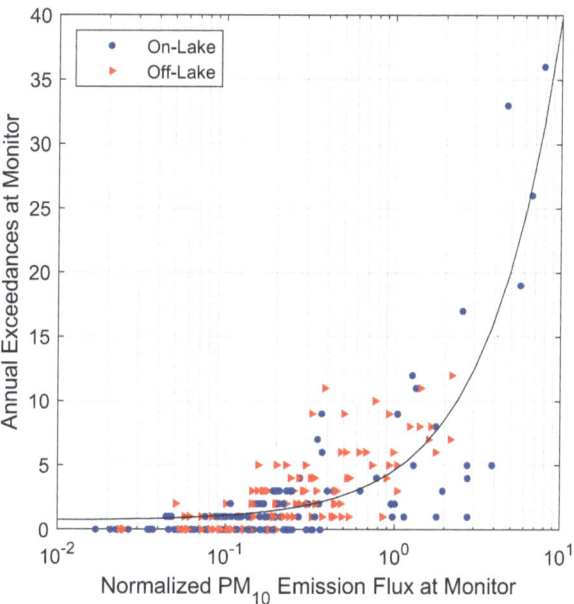

FIGURE A-2 Relationship between normalized emission flux and annual exceedances for all monitoring sites between 2000 and 2023.

using the wind direction sectors specified by the District. They do not include exceedances that are attributed partially to both on- and off-lake sources. The plot shows that, as expected, the number of annual exceedances at a monitor increases with upwind emission flux. Note that relative emission flux (normalized to Keeler Dunes) as low as 0.2 can lead to 5 exceedances at a monitor.

Although all available data from 2000 to 2023 from all monitors was used for Figure A-2, Keeler and Dirty Socks were the only monitoring stations for which the panel had complete wind and PM_{10} concentration records from 2000 to 2023 (Figures A-3 and A-4). The panel did not have access to complete hourly wind and concentration data for the other monitors until after 2008, and many monitoring stations were also missing additional data between 2008 and 2023. Trends for the data at the Keeler and Dirty Socks monitors were extracted with singular spectrum analysis (SSA; Elsner and Tsonis 1996; Vautard, Yiou, and Ghil 1992), a decomposition technique widely used in climatology and meteorology.

The shaded areas around the trend lines are the 95 percent confidence intervals of the trends obtained by bootstrapping the residuals between the computed fluxes and the first estimate of the trend from the SSA. The computed fluxes (based on measured concentrations and wind speeds) are assumed to be lognormally distributed about the corresponding SSA prediction so that the residual is the logarithm of the computed flux to the corresponding SSA estimate. A set of 1,000 "pseudo-fluxes" were created by adding the residuals randomly to the logarithms of the SSA estimates. Then, SSA trends were created for each of the pseudo-fluxes. The limits of the shaded areas for each year correspond to the 2.5th and the 97.5th percentiles of the 1,000 SSA predictions for each year.

At Keeler and Dirty Socks, current off-lake emission fluxes are markedly higher than on-lake fluxes and show small or no statistically significant temporal decline (Figures A3 and A4). Given these sustained off-lake levels, exceedances of the PM_{10} standard at downwind monitors are likely unless additional controls are implemented.

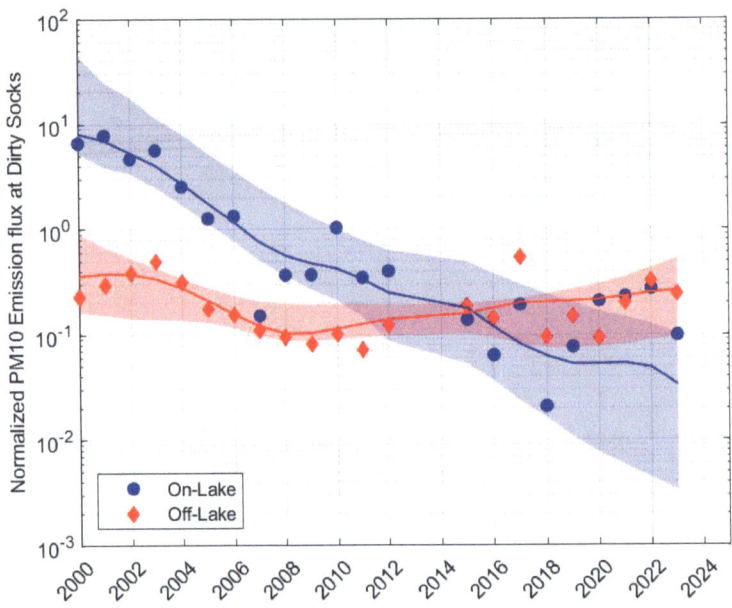

FIGURE A-3 Trends in on- and off-lake normalized PM_{10} emissions at Dirty Socks.
NOTES: The markers are the computed average normalized emission fluxes. The solid lines passing through these points are the mean trends modeled with SSA. The shaded areas around the trend lines are the 95 percent confidence limits of the trends. Mann-Kendall tests show a statistically significant decreasing trend for on-lake sources (p-value <0.01), and no statistically significant trend for off-lake sources (p-value of 0.96).

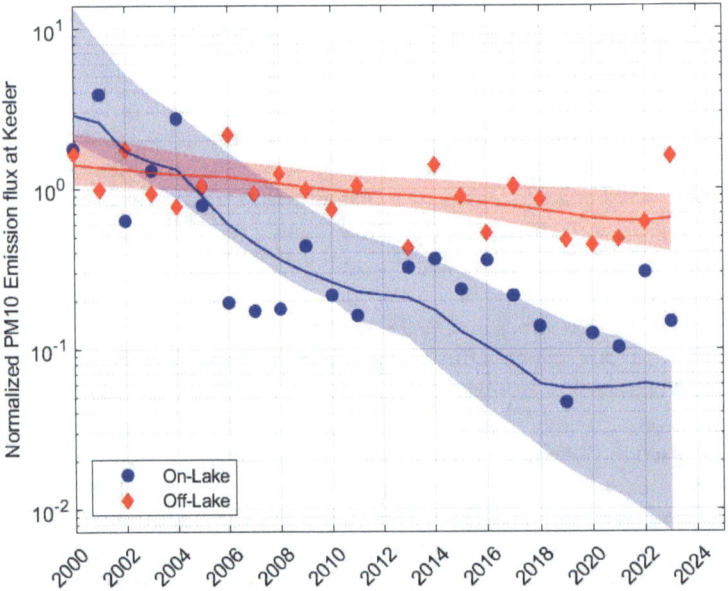

FIGURE A-4 Trends in on- and off-lake normalized emissions at Keeler.
NOTES: The markers are the computed average normalized emission fluxes. The solid lines passing through these points are the mean trends modeled with SSA. The shaded areas around the trend lines are the 95 percent confidence limits of the trends. Mann-Kendall tests show a steep, statistically significant decreasing trend for on-lake sources (p-value <0.01), as well as a more shallow, statistically significant decreasing trend for off-lake sources (p-value <0.01).

Appendix B

Advisory Panel Biographical Sketches

ARMISTEAD (TED) RUSSELL is the Howard T. Tellepsen Chair and Regents' Professor of Civil and Environmental Engineering at Georgia Institute of Technology, where his research is aimed at better understanding the dynamics of air pollutants across scales and assessing their impacts on health and the environment to develop approaches to design strategies to effectively improve air quality. Among Dr. Russell's honors are the Federal Highway Administration Environmental Excellence Award for work on near-road impact and being elected to the Canadian Academy of Engineering. He was a member of the Environmental Protection Agency's Clean Air Science Advisory Committee (CASAC) and a member of the National Research Council's Board on Environmental Studies and Toxicology. Dr. Russell chaired the CASAC NOx-SOx, Secondary National Ambient Air Quality Standards review panel, the Ambient Air Monitoring Methods Subcommittee, and the Council on Clean Air Compliance Analysis' Air Quality Modeling Subcommittee. He has served on and chaired multiple National Academies of Sciences, Engineering, and Medicine committees, including the committee for Assessing Causality from a Multidisciplinary Evidence Base for National Ambient Air Quality Standards and the first task of the Owen's Lake Scientific Advisory Panel. Dr. Russell earned his B.S. from Washington State University and his M.S. and Ph.D. degrees at the California Institute of Technology, conducting his research at Caltech's Environmental Quality Laboratory.

SARAH AARONS is an assistant professor at Scripps Institution of Oceanography. Her research focuses on the relationship between climate and the Earth's surface through the lens of isotope geochemistry, with a strong focus on the mineral dust cycle. Dr. Aarons's research tracks variations in the sources of dust to the cryosphere and mountain ecosystems both in the past and the modern day. She is a recipient of the F. W. Clarke Award from the Geochemical Society and the Doris M. Curtis Award from the Geological Society of America. Dr. Aarons received her B.S. degree in geological and environmental sciences from Stanford University and her Ph.D. in geology from the University of Michigan.

ROYA BAHREINI is a professor of atmospheric science in the Department of Environmental Sciences at University of California (UC), Riverside. She specializes in airborne, ground-based, and laboratory measurements of aerosol composition and microphysical properties to understand aerosol sources and formation process, influence on air quality, and direct and indirect effects on climate. Before joining UC Riverside, Dr. Bahreini was a Cooperative Institute for Research in Environmental Sciences (CIRES) visiting postdoctoral fellow at University of Colorado Boulder (2005–2007), a research scientist at CIRES and the National Oceanic and Atmospheric Administration (NOAA) Earth System Research Laboratories (ESRL; 2007–2012), and University of Denver (2012). She is a recipient of the National Science Foundation CAREER Award, the Thomson Reuters Highly Cited Researchers Award (2014), as well as The World's Most Influential Scientific Minds Award (2014). Dr. Bahreini served on the National Academies of Sciences, Engineering, and Medicine's first task for the Owens Lake Scientific Advisory Panel. She currently serves on the South Coast Air Quality Management District's Technical Advisory Group for the Multiple Air Toxics Exposure Study VI (MATES VI) and California Air Resources Board's Research Screening Committee. Dr. Bahreini received her B.S. in chemical engineering from University of Maryland, College Park (1999), and her M.S. (2003) and Ph.D. (2005) degrees in environmental science and engineering from the California Institute of Technology.

DAVID DuBOIS is the State Climatologist and associate college professor in the Department of Plant and Environmental Sciences at New Mexico State University. Prior to 2010, he was employed as a scientist and a graduate student at the Desert Research Institute, and was manager at the State of New Mexico's Air Quality Bureau. Dr. DuBois is the first contact person for climate information in New Mexico and oversees the operations, maintenance, and data delivery from the statewide ZiaMet mesonet. He is involved in science, technology, engineering, and mathematics (STEM) outreach, citizen science, and climate literacy programs. Dr. DuBois chairs the New Mexico Drought Monitoring Workgroup that tracks the status of the drought in New Mexico, and he conducts research in air quality and climate. In 2023, he shared the John C. Frye Memorial Award in Environmental Geology for Bulletin 164 with 11 other colleagues. Dr. DuBois was awarded the Community Leader Public Health Hero Award and the Climate Direct Action Award by the Climate Change Leadership Institute. Dr. DuBois holds a B.S. and M.S. in physics and a Ph.D. in atmospheric sciences from the University of Nevada, Reno.

VALERIE EVINER is a professor of ecosystem management and restoration in the Department of Plant Sciences and an ecologist in the Agriculture Experiment Station at the University of California, Davis. Her research focuses on developing a mechanistic understanding of plant-soil interactions to increase the understanding, restoration, and effective management of ecosystem services, plant invasions, plant communities, and ecosystem resilience in response to multiple environmental changes. Dr. Eviner's current projects include understanding the building blocks of resilience in California's ecosystems, how those vary in relation to multiple changes (including drought, wildfire, and invasion), and how those can be managed under changing conditions. She is a fellow of the Ecological Society of America, the American Association for the Advancement of Science, and the Earth Leadership Program. Dr. Eviner is on the scientific advisory boards of the California Climate and Agriculture Network, Point Blue Conservation Science, and the California Native Grasslands Association, and she is on the editorial board of *Ecosystems*. Dr. Eviner served on the National Academies of Sciences, Engineering, and Medicine's first task for the Owens Lake Scientific Advisory Panel. She received a B.A. in biology from Rutgers University and a Ph.D. in integrative biology from the University of California at Berkeley.

SHANNON MAHAN is the director the U.S. Geological Survey (USGS) Luminescence Geochronology Lab. She began work for the USGS in Denver, Colorado, in 1987 and eventually wound her way through four teams/branches/science centers before landing with the Environmental and Geosciences Science Center in 2015. Mahan has been working in the science of luminescence geochronology and radiation dosimetry since 1990. She has presented more than 300 diverse studies relating to luminescence dating at national and international scientific forums, authored or coauthored more than 200 peer-reviewed articles or maps, and supervised the luminescence studies for more than 50 graduate and post-graduate students. Mahan is frequently asked to review papers for prestigious quaternary science journals, perform guest editor duties, review National Science Foundation grants, and provide dating control for important archeological, paleontological, and geological sites. She is a fellow of the Geological Society of America and a member of the Association for Women Geoscientists and Federally Employed Women. Mahan received her B.S. from Adams State University in Colorado.

TOM MOORE coordinates air quality planning for the Denver Metro/North Front Range–Regional Air Quality Council (RAQC). He has also worked for the Colorado Air Pollution Control Division after managing the Western Regional Air Partnership organization and technical project activities for more than 20 years, most recently at the Western States Air Resources Council (WESTAR) and previously at the Western Governors' Association. Moore has led numerous air pollution studies and analysis projects, held management positions in state and local governments, and has worked as an environmental consultant. Before his regional role at the Western Regional Air Partnership (WRAP), he managed air quality monitoring and analysis activities for the Arizona Department of Environmental Quality, where he assisted in the development and led the implementation of particulate matter health and visibility monitoring networks throughout the state. Moore has also served on federal Clean Air Act national advisory groups for air quality health and welfare standards (such as the Clean Air Scientific Advisory Committee Particulate Matter Panel, 2008–2011) and regional haze (including ozone/particulate matter/relative humidity implementation under the Federal Advisory Committee Act). He received a B.S. in physical geography from Arizona State University.

GREGORY OKIN currently serves as the chair of geography at University of California, Los Angeles (UCLA), and holds an additional appointment in UCLA's Institute of Environment and Sustainability. His research interests center on drylands including plant-soil interactions, surface-atmosphere interactions, aeolian processes and dust production, and remote sensing. Dr. Okin is a fellow of the American Geophysical Union and served on the National Academies of Sciences, Engineering, and Medicine's first task for the Owens Lake Scientific Advisory Panel. He received a B.A. with a double major in philosophy and chemistry from Middlebury College and a Ph.D. from California Institute of Technology (Caltech) in geological and planetary sciences.

DANI OR is a Nevada Engineering Distinguished Professor in the Department of Civil and Environmental Engineering at the University of Nevada, Reno. Previously he was a professor of terrestrial environmental physics at the Swiss Institute of Technology Zurich. Dr. Or's research focuses on mass and energy transport in porous media; mechanics of landslides and avalanches; evaporation from porous surfaces; biophysical processes and biological activity in soil; and more recently, the physical impacts of wildfire on soil properties and functioning. He also works on surface evaporation from complex surfaces, turbulences and mechanics of boundary layers, capillary forces in porous media, and development of biocrusts. Dr. Or is the recipient of the Kirkham Soil Physics Award (2001), and he was 2004 fellow of the Soil Science Society of America, chair of the 2008 Gordon Research Conference on Flow and Transport (Oxford, UK), and 2010 fellow of the American Geophysical Union. He was recipient of the 2013 Helmholtz International Fellow Award, 2014 elected fellow of the Geological Society of America, and was awarded the 2017 European Geosciences Union John Dalton Medal. Dr. Or presented the 2018 Langbein lecture of the American Geophysical Union, and he was elected to the National Academy of Engineering in 2022. He was selected Caltech Moore Scholar in 2023. Dr. Or holds bachelor's and master's degrees from the Hebrew University of Jerusalem and a Ph.D. in soil and biometeorology (soil physics) from Utah State University.

ROBERT SCOTT VAN PELT has served as a research soil scientist with the U.S. Department of Agriculture (USDA) Agricultural Research Service's (ARS's) Wind Erosion and Water Conservation Research Unit in Big Spring, Texas, for the past 24 years. Prior to his current position, he was employed as a consultant with AgPro, LLC of Carlsbad, New Mexico, INTERA, Inc. of Austin, Texas, and Soil and Water West, LLC of Rio Rancho, New Mexico. Dr. Van Pelt is most noted for his expertise in wind erosion processes; dust generation; dust physical, chemical, and microbiological characteristics; and dust transport. He is a member of Phi Kappa Phi National Honor Society and a member of the Soil and Water Conservation Society for which he serves as an associate editor of the *Journal of Soil and Water Conservation*. Dr. Van Pelt received the Conservation Research Award in 2015 and the Associate Editor Award of Excellence in 2018 and is a founding member of the International Society for Aeolian Research for which he has served on the board of directors. He is an associate editor of *Aeolian Research* and is a past president of the International Soil Conservation Organization, where currently he serves on the board of directors. Dr. Van Pelt is also a founding member of the Dust Alliance for North America, currently serving as a member of the executive committee and as president. He served on the National Academies of Sciences,

Engineering, and Medicine's first task for the Owens Lake Scientific Advisory Panel. Dr. Van Pelt received his B.S. in biology from the University of New Mexico in 1978, his M.S. in floristics and plant ecology from the University of New Mexico in 1984, and his Ph.D. in soil physics from New Mexico State University in 1990.

AKULA VENKATRAM is a distinguished professor of mechanical engineering at the University of California, Riverside. His research is focused on the development and the application of models for the transport and dispersion of air pollutants over urban and regional scales. Dr. Venkatram coedited and contributed to the "Lectures on Air Pollution Modeling" published by the American Meteorological Society. He was member of the team that developed AERMOD, the primary regulatory model recommended by the U.S. Environmental Protection Agency for estimating the impact of pollution sources. Dr. Venkatram is the recipient of the inaugural award from the American Mathematical Society's Committee on Meteorological Aspects of Air Pollution for "contributions to the field of air pollution meteorology through the development of simple models in acid deposition, ozone photochemistry and urban dispersion." His research on small-scale dispersion modeling is summarized in the monograph *Urban Transportation and Air Pollution*, published by Elsevier. Dr. Venkatram served on the National Academies of Sciences, Engineering, and Medicine's first task for the Owens Lake Scientific Advisory Panel. He received his Ph.D. in mechanical engineering from Purdue University.

IAN WALKER is a professor of geography at the University of California, Santa Barbara. Dr. Walker's expertise includes aeolian and coastal geomorphology, sediment transport and erosion processes, environmental fluid dynamics, sand dune systems, dune restoration, and dust mitigation strategy development. His research uses close range remote sensing (LiDAR, unmanned aerial systems), sedimentology, modeling, and field experiments to better understand and manage dynamic landscapes. Dr. Walker is also experienced with ecosystem restoration and mitigation of dust emissions and sand drift hazards in aeolian landscapes. He has engaged in several dune restoration projects and is an appointed member of the Oceano Dunes Scientific Advisory Group, established in 2018 per a Stipulated Order of Abatement issued by the San Luis Obispo County Air Pollution Control District against the California Department of Parks and Recreation. As such, Dr. Walker has appreciable experience working within environmental management and air quality regulatory frameworks and with air quality regulators at county and state levels to understand and help address regulatory issues, exceptional events, and dust mitigation strategies toward achieving standards while balancing competing interests. He obtained his B.S. in physical geography from the University of Toronto and his Ph.D. from the University of Guelph in Canada.

Appendix C

Glossary

Aeolian/eolian: "Pertaining to the wind; [especially] said of such deposits as loess and dune sand, of sedimentary structures such as wind-formed ripple marks, or of erosion and deposition accomplished by the wind." (American Geological Institute 1983)

Alluvial: "Pertaining to or composed of alluvium, or deposited by a stream or running water." (American Geological Institute 1983)

Alluvial fan: "An outspread, gently sloping mass of alluvium deposited by a stream, [especially] in an arid or semiarid region where a stream issues from a narrow canyon onto a plain or valley floor. Viewed from above, it has the shape of an open fan, the apex being at the valley mouth." (American Geological Institute 1983)

Bajada: "A broad, gently inclined detrital surface extending from the base of mountain ranges out into an inland basin, formed by the lateral coalescence of a series of alluvial fans and having an undulating character due to the convexities of the components fans." (American Geological Institute 1983)

Colluvium: "A general term applied to loose and incoherent deposits, usually at the foot of a slope or cliff and brought there chiefly by gravity." (American Geological Institute 1983)

Design value: "A statistic that describes the air quality status of a given location relative to the level of the National Ambient Air Quality Standards (NAAQS)." (EPA 2025a)

Dune: "A mount, ridge, or hill of wind-blown sand, either bare or covered with vegetation." (American Geological Institute 1983)

Fluvial: "Of or pertaining to rivers." (American Geological Institute 1983)

Histic epipedon: "A surface soil horizon, not less than 1 m in depth, high in organic carbon, and saturated with water for some part of the year." (Oxford Reference 2015)

Interfluve: "The relatively undissected upland between adjacent streams flowing the same general direction." (American Geological Institute 1983)

Lacustrine: "Pertaining to, produced by, or inhabiting a lake or lakes." (American Geological Institute 1983)

Local off-lake sources: Emissive sources above the 3,600-ft regulatory shoreline at Owens Lake, but within the Owen Valley Planning District.

Off-lake sources: Emissive sources above the 3,600-ft regulatory shoreline at Owens Lake.

On-lake: Emissive sources below the 3,600-ft regulatory shoreline at Owens Lake.

Playa: "A term used in the southwestern U.S. for a dry, barren area in the lowest part of an undrained desert basin, underlain by clay, silt, or sand, and commonly by soluble salts." (American Geological Institute 1983)

Rindge soils: "Very deep, very poorly drained organic soils that formed in fresh water marshes, sloughs and drainage channels from mixed decomposed reeds, tules and alluvium." (USDA 2016)

Sand: "A detrital particle smaller than a granule and larger than a silt grain, having a diameter in the range of 1/16 to 2 mm." (American Geological Institute 1983)

Sand sheet: "A large irregularly shaped plain of eolian sand, lacking the discernible slip faces that are common on dunes." (USGS 2024)

Saltation: "Sediment transport in which particles are moved forward in a series of short leaps or bounces." (American Geological Institute 1983)

Silt: "A detrital particle finer than fine sand and coarser than clay, commonly in the range of 1/16 to 1/256 mm." (American Geological Institute 1983)

Suspension: "A mode of sediment transport in which the upward currents in eddies of turbulent flow are capable of supporting the weight of sediment particles and keeping them indefinitely held in the surrounding fluid." (American Geological Institute 1983)

Winnowing: "Separation of fine particles from coarser ones by action of the wind." (American Geological Institute 1983)